Quantum Computing Solutions

Solving Real-World Problems Using Quantum Computing and Algorithms

Bhagvan Kommadi

Apress®

Quantum Computing Solutions: Solving Real-World Problems Using Quantum Computing and Algorithms

Bhagvan Kommadi
Hyderabad, India

ISBN-13 (pbk): 978-1-4842-6515-4
https://doi.org/10.1007/978-1-4842-6516-1

ISBN-13 (electronic): 978-1-4842-6516-1

Managing Director, Apress Media LLC: Welmoed Spahr
Acquisitions Editor: Susan McDermott
Development Editor: Laura Berendson
Coordinating Editor: Jessica Vakili

Distributed to the book trade worldwide by Springer Science+Business Media New York, 1 NY Plaza, New York NY 10004. Phone 1-800-SPRINGER, fax (201) 348-4505, e-mail orders-ny@springer-sbm.com, or visit www.springeronline.com. Apress Media, LLC is a California LLC and the sole member (owner) is Springer Science + Business Media Finance Inc (SSBM Finance Inc). SSBM Finance Inc is a **Delaware** corporation.

For information on translations, please e-mail booktranslations@springernature.com; for reprint, paperback, or audio rights, please e-mail bookpermissions@springernature.com.

Apress titles may be purchased in bulk for academic, corporate, or promotional use. eBook versions and licenses are also available for most titles. For more information, reference our Print and eBook Bulk Sales web page at www.apress.com/bulk-sales.

Any source code or other supplementary material referenced by the author in this book is available to readers on GitHub via the book's product page, located at www.apress.com/978-1-4842-6515-4. For more detailed information, please visit www.apress.com/source-code.

Printed on acid-free paper

*Writing a book on quantum computing was harder than
I thought and more rewarding than I could have ever imagined.
None of this would have been possible without support from
my parents.*

Table of Contents

About the Author

Bhagvan Kommadi is the founder of Architect Corner, an AI startup, and he has 20 years of industry experience ranging from large-scale enterprise development to helping incubate software product startups. He has a master's degree in industrial systems engineering from Georgia Institute of Technology and a bachelor's degree in aerospace engineering from the Indian Institute of Technology, Madras. He is a member of the IFX forum, a member of Oracle JCP, and a participant in Java Community Process.

Bhagvan founded Quantica Computacao, the first quantum computing startup in India. He has engineered and developed simulators and tools in quantum technology using IBM Q, Microsoft Q#, and Google QScript. The company's focus is developing quantum cryptographic tools to provide quantum proof data security, which will help banking institutions protect their transactions. He is now the director of product engineering at Value Momentum. Value Momentum offers a social network for doctors (White Coats) and provides telehealth support through the Practice Plus suite of products and services.

Bhagvan has published papers and presented at IEEE, AstriCon, Avios, DevCon, PyCon, and genomics and biotechnology conferences on topics including adaptive learning, AI Coder, and more. He is a hands-on CTO who has contributed to open source, blogs, and the latest technologies such as Go, Python, Django, node.js, Java, MySQL, Postgres, Mongo, and Cassandra. He was the technical reviewer for the book *Machine Learning with TensorFlow* and is the author of *Hands-on Data Structures and Algorithms with Go and Paytech* and *The Payment Technology Handbook for Investors, Entrepreneurs, and FinTech Visionaries.*

About the Technical Reviewer

Mr. Nixon Patel is a visionary leader, an exemplary technocrat, and a successful serial entrepreneur with a proven track record for growing 7+ businesses from startup to millions in annual sales and 1,000+ employees with large technology and operations organizations across the globe in the quantum computing, artificial intelligence, deep learning, big data analytics, cloud, IOT, speech recognition, machine learning, renewable energy, information technology, telecom, and pharmaceutical industries, all in a short span of 30 years. Most recently he exited an AI and deep learning company with a 5X valuation; before that he was the chief data scientist at Brillio, a sister company of Collabera, where he was instrumental in starting the big data analytics, cloud, and IOT practices. He has a bachelor's of technology degree (hons) in chemical engineering from IIT Kharagpur, a master's of science in computer science from New Jersey Institute of Technology, a data science specialization from Johns Hopkins University, and a certificate in quantum computing and information from the Massachusetts Institute of Technology. He is currently pursuing a second master's in business and science in analytics from Rutgers University. Currently he holds the positions of adjunct professor and mentor at IIT Hyderabad and NSHM College Kolkata.

PART I

Introduction

CHAPTER 1

Quantum Solutions Overview

Introduction

Quantum computation is...a distinctively new way of harnessing nature....
It will be the first technology that allows useful tasks to be performed in
collaboration between parallel universes.

—David Deutsch

This chapter gives an overview of quantum solutions. You will see how quantum computing can be applied in real life. The problem and solutions are presented with the solution benefits. The solution benefits are automation, cost reduction and profit improvements, efficiency improvement, and defect reduction.

The solutions covered in this chapter are related to cryptography, optimization, and cybersecurity.

Let's start by looking at the history of quantum computing and key achievements so far.

Richard Feynman proposed quantum electrodynamics in 1965 (see Figure 1-1). Quantum electrodynamics is related to the interaction of the electrons using the electromagnetic force of the photon. He invented the particles named *antiparticles*. Antiparticles are the particles that have charges opposite to the mirror particles.

© Bhagvan Kommadi 2020
B. Kommadi, *Quantum Computing Solutions*, https://doi.org/10.1007/978-1-4842-6516-1_1

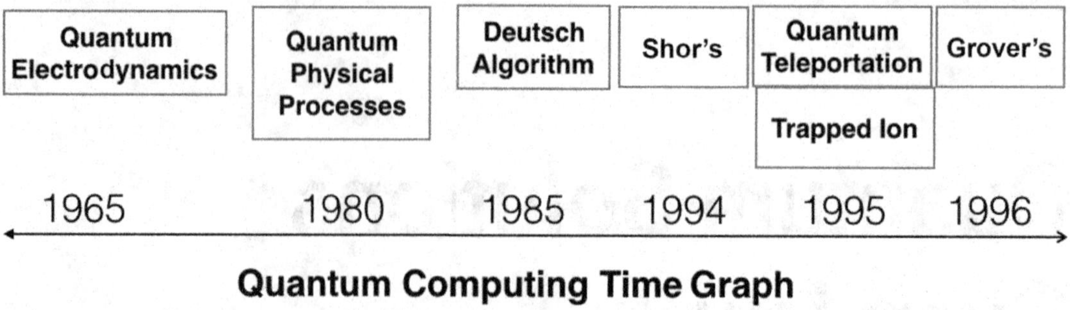

Quantum Electrodynamics	Quantum Physical Processes	Deutsch Algorithm	Shor's	Quantum Teleportation	Grover's
				Trapped Ion	

1965	1980	1985	1994	1995	1996

Quantum Computing Time Graph

Figure 1-1. *Quantum computing time graph*

He came up with new concepts related to quantum physical processes in 1980. He proposed that quantum states are represented by binary numbers.

David Deutsch came up with a concept called *universal quantum computers* in 1985. A universal quantum computer is based on the two-state system, which consists of a set of quantum gates. *Quantum gates* are a set of basic operations.

Peter Shor invented the quantum entanglement–based method for prime number factorization in 1994. This method is used to identify the prime factors of big numbers. The efficiency of the algorithm was much better compared to the previous methods. This research helped in creating research interest at other universities.

The trapped ion concept was first developed by the National Institute of Standards and Technology (NIST) and the California Institute of Technology in 1995. The concept is related to the ions that are trapped, and the temperature is decreased to a quantum state. In 1996 Bell Laboratories scientist Loy Grover developed Grover's algorithm, which is related to searching using quantum algorithms. A research team consisting of members from the University of California, Massachusetts Institute of Technology (MIT), Harvard University, and IBM invented a nuclear magnetic-resonance technology to modify quantum information in a liquid state.

David DiVincenzo proposed the basic characteristics for real quantum computer technology. Before David DiVincenzo, while Seth Lloyd was working in Los Alamos National Laboratory, he invented the molecular quantum computer. His proposal was based on a chain molecule structure. This structure has a pattern of little blocks, and it is repeating (see Figure 1-2).

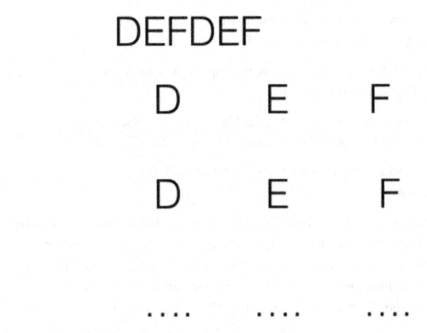

Figure 1-2. *Chain molecule structure*

This proposal was the basis for the first quantum computer concept. Seth Lloyd's proposal was not created, but in 2000, David DiVincenzo created a list of criteria for quantum computer technology. The criteria are related to quantum computing and communication models.

- A quantum computer consists of well-characterized quantum bits, and it is a scalable physical system.

- It needs to have the capability to create a set of quantum bits.

- A quantum computer can set the initial state of the quantum bits. This is similar to a reset button on a desktop computer.

- A quantum computer has long decoherence times.

- Quantum computer has error correction related to the data retention in memory.

- A quantum computer is based on a universal set of quantum gates. Quantum algorithms need to be executed on the computers. These algorithms might involve the rotation of quantum bits on a Bloch sphere and the entanglement of qubits.

- It needs to have a measurement feature related to the state of quantum bits.

- A quantum computer needs to have the capability to interchange the static and dynamic quantum bits.

- Information needs to be transferred along large distances. Quantum bits need to be transmitted between the source and destination.

- The information needs to be retained in quantum bits after transmission.

After DiVincenzo, Van Meter presented his thesis for building a quantum computer. The technology evolved from Van Meter's thesis to a physical quantum computer. D-Wave came up with its own quantum computer. The latest update is that IBM is building a 53 quantum bit quantum computer. Google already has a 72 quantum bit quantum computer. The IBM Quantum computer will be accessible through its IBM Bluemix cloud environment.

Real-Life Problems and Solutions

Quantum solutions are related to building quantum algorithms for improving computational tasks within quantum computing, artificial intelligence, data science, and machine learning. Quantum solutions are picking up speed compared to quantum computer innovations. In this chapter, we look at various solutions such as optimization problems, quantum cryptography, and cybersecurity. This chapter explains the quantum solutions involving AI algorithms and applications in different areas. The other chapters of this book will present code samples. They will be based on real-life problems such as risk assessment and fraud detection in banking. In pharma, we look at drug discovery and protein folding solutions. Supply chain optimization and purchasing solutions are presented in the manufacturing domain. In the utility industry, energy distribution and optimization problems are explained in detail with solutions. Advertising scheduling and revenue optimization solutions are covered in the media and technology verticals.

The areas in which quantum machine learning are being applied are nanoparticles, material discovery, chemical design, drug design, pattern recognition, and classification. The applicable use cases are creating new materials that can be applied in space tech, wearable tech, renewable energy, nanotech, new drugs and chemical combinations, genetic science, biometrics, IoT devices, and universe discovery.

Note   Rose's law states that the computational power of quantum computers doubles every 12 months, as opposed to the 18 months doubling of Moore's law.

"Quantum computers have the potential to solve problems that would take a classical computer longer than the age of the universe."

—David Deutsch, Oxford University professor

Barclays, Goldman Sachs, and other financial firms are investigating the potential use of quantum computing in areas such as portfolio optimization, asset pricing, capital project budgeting, and data security. In aerospace, Airbus is exploring applications in communications and cryptography.

Lockheed Martin is investigating applications of verifying and validating complex systems. The US Navy has plans to develop algorithms for optimization problems such as data storage and energy-efficient data retrieval with underwater autonomous robots. NASA is exploring applications in communications, distributed navigation, and system diagnostics. Technology players such as Alibaba, Google, and IBM are working on applications such as hack-resistant cryptography, software debugging, and machine learning. Life sciences firms are seeking applications of quantum computing in personalized medicine and drug discovery.

The following list presents use cases from different domains. These use cases prove that quantum computing can help human life by using quantum technology and its high-performance processing features.

- Economy and finance (see Figure 1-3)

Figure 1-3. *Stock market*

- Analysis and simulation of stock portfolios for investment decisions

- More effective fraud detection and risk optimization in real time

- Medicine (see Figure 1-4)

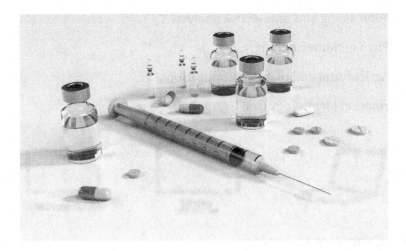

Figure 1-4. *Medical sciences*

- Research and development of medical sciences (e.g., analysis of DNA sequences that are usually large)

- Bioengineering and telemedicine relying on heavy image processing and analysis

- Faster and more accurate diagnosis of critical diseases such as cancer

- Drug discovery and production at scale

- Natural sciences (see Figure 1-5)

Figure 1-5. *Natural sciences*

- Forecasting and time-series analysis

- Plant genome analysis

- Agricultural and planting applications

- Information technology (see Figure 1-6)

Figure 1-6. *Information technology*

- Data searches and analysis of big data warehouses

- Software tests and simulations that are too complex and time-consuming for current computers

- Management optimization (see Figure 1-7)

Optimization

Figure 1-7. *Optimization*

- Supply chain and procurement (scenario analysis and decision-making with regard to space and cost efficiency, distribution channels, transportation optimization, etc.)

- Asset and resource management (probabilistic modeling for cost savings and lifecycle management)

- Analytical capabilities in general, which leads to more insightful decision-making process

Solution Benefits

Quantum solution infrastructure typically offers a high degree of automation and delivers significantly improved performance and resilience. A platform powered by a quantum computing framework evolves through continuous learning and deep learning. During the evolution, the system becomes a self-reliant and self-sustained platform. The platform will be able to make critical decisions.

This new paradigm of computing will change the way software is written. The framework evolved from basic programming software to patterns and models. This paradigm will change the deployment and management of the applications. These solutions are based on deep learning–based frameworks, quantum computing platforms, and a new set of tools. These quantum solution frameworks will enable organizations to create, test, and deploy new applications without any human touch points. The development cycle will help in rebuilding the software with deep learning and quantum computing algorithms.

The infrastructure for a quantum solution needs to be highly scalable. The quantum solutions platform requires the high-performance computing power to be able to process data loads and guarantee service assurance and performance. Current applications lack performance and service levels on a millisecond basis and millions of cycles per second. High-quality quantum solutions will be built with quantum algorithms and quantum AI techniques. New quantum solutions infrastructure will use the GPU-level computing power blended with quantum computing. This infrastructure will leverage open source software and the cloud for elastic capacity demand.

Note   The quantum cloud consists of quantum simulators and processing applications. Quantum cloud services are provided by IBM, Google, Rigetti, and Microsoft for information processing.

Quantum solutions infrastructure will be IoT enabled. Data connectivity and real-time insight are important factors for real-time quantum solutions. This solution infrastructure will have features to secure data transfer, ingestion, and data processing. The data will be acquired from various data sources to create real-time insights. IoT solution deployments will have all connected devices. A policy and governance framework will be part of quantum solutions.

Quantum solution infrastructure needs to be secure. Enterprises will have quantum solutions that involve cybersecurity for the infrastructure. The servers need to be monitored, and the security data needs to be ingested, integrated, and stored for analysis. The security and monitoring events need to be filtered using unsupervised learning. The threat intelligence model is built and has people, places, and applications deployed on the server. Prevention of the threats is possible through an event dashboard.

Automating Manual and Semi-manual Processes

Quantum solutions involve the automation of business processes. Automation helps in enhancing the businesses and enterprises by changing the way they operate. Quantum solutions involve artificial intelligence, machine learning, robotic process automation, and business process automation methods. These show immediate benefits in enterprises. See Figure 1-8.

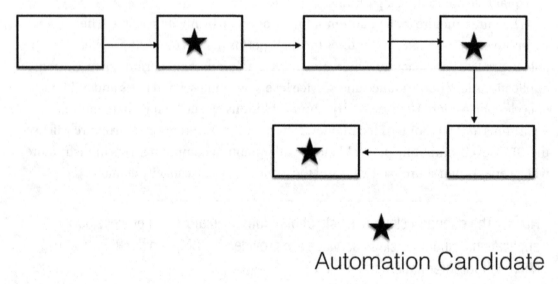

Figure 1-8. *Process Automation*

Handling big data is the biggest automation effort that quantum solutions can provide with their high-performant data processing capabilities. Human errors and problems arise when adding automation to the business processes. Predictive and prescriptive analytics are solutions that can add value to the enterprise business's decision-making.

The data comes from the different servers, applications, devices, and departments in a business organization. The data needs to be stored, cleaned, and analyzed for actionable decisions. Quantum solutions help in handling big data and creating dashboards and reports.

Reducing Costs and Improving Profits

The effort and cost for deployment and manageability need to be balanced with the quality of the solution. Having fewer resources, minimizing the operational costs, and delivering high-quality solutions are the key goals for enterprise IT. Quantum solutions help cut down the infrastructure costs with exponential speedup in the algorithms compared to the classical algorithms. See Figure 1-9.

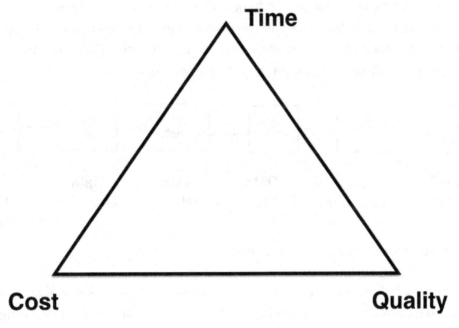

Figure 1-9. *Triangle showing the IT challenges*

Many enterprises have challenges related to resources dedicated to the maintenance of the applications (80 percent of them). The operating costs of the deployed solutions are twice the solution development or package software cost. The developer resource costs are more than the software servers' cost by a ratio of 18:1. Banks are spending about a billion dollars per year to maintain their applications and solutions that are up and running 24 hours every day of the week. This amounts to the total cost of ownership and challenges for maintaining IT in a bank.

The other challenges are related to legacy apps and products, outdated technology platforms, and not having skilled resources in the IT departments. Time to market is an important factor in every aspect of the business. Technology platforms do not have the capabilities to match the speed expected in terms of time to market. The volume and complexity of change or requirements do not match the resources available in the enterprise.

Having effective solutions is important to handle this exponential increase in the cost of maintaining the deployed solutions. Quantum solutions solve this problem by cutting the resources required by offering an exponential speedup in the algorithms.

The quantum solutions infrastructure will be highly performant, secure, available, and resilient, and it will be able to match the time to market needs. The people needed will require average skills but will be able to deploy these solutions easily and upgrade them for maintenance. Overall, enterprises will be able to handle their customer requirements quickly and develop products faster. See Figure 1-10.

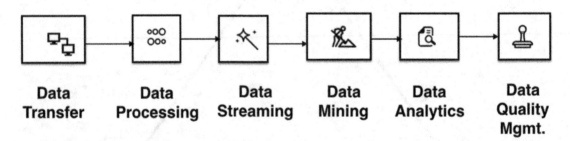

Figure 1-10. *Data management process*

Quantum solution infrastructure will have data models, patterns, and algorithms to process data quickly. First, The data can be from various sources, and processing power helps in creating reports. Decision-making will be faster using the reports and dashboards. The total cost of ownership will go down for an enterprise.

Second, this complexity is a significant limiting factor for executing product development at the speed necessary to respond to market needs. In the absence of careful planning, technology platforms will fail to anticipate the complexity of interactions. Managing volume and complexity falls to a highly skilled human workforce, the demand for which is already far outstripping supply.

Without effective solutions, the cost of maintaining applications will, therefore, increase exponentially. Furthermore, increasing complexity means a greater risk of security loopholes, making organizations more vulnerable to threats.

Improving Efficiencies and Reducing the Defects

The manageability of the solution infrastructure is the key for quantum solutions. There will be an impact to the cost and security of the solution infrastructure because of the complexity in handling and managing the applications. Visibility and monitoring of the applications deployed are achievable with threat intelligence models using deep learning. New deployments, upgrades, and releases can be handled with change and impact analysis. This analysis helps in identifying IT problems and maintaining the deployed applications without issues and errors. Defects cost 100 times more to fix in the production stage compared to the design stage. See Figure 1-11.

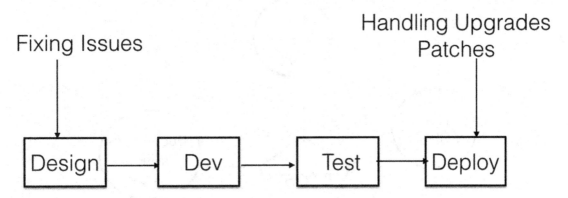

Figure 1-11. *Solution development process*

Solutions

In this section, we look at different quantum solutions in the areas of cryptography, optimization, and cybersecurity.

Cryptography

Banks will be interested in protecting their payment gateways and solutions against security threats from future quantum computing capabilities. There is a need to make the payment solutions "quantum-resistant." In addition, there is a need for blockchain-based banking and financial solutions to improve using post-quantum cryptographic algorithms.

The Quantum Resistant Ledger is the next-generation solution for banking and financial services. Quantum cryptography is based on quantum key distribution. Quantum key distribution is related to the expansion of a short, shared key into an infinite shared stream. This demands improvement in the efficiency of post-quantum cryptographic algorithms.

Financial firms need to start investing in post-quantum cryptography. Blockchain-based solutions will be not quantum-resistant if they are not upgraded to post-quantum cryptographic algorithms. Functioning cryptographic systems are DES, Triple DES, AES, RSA, Merkle hash-tree signatures, Merkle–Hellman knapsack encryption, Buchmann–Williams class-group encryption, ECDSA, HFEv-, and others (see Figure 1-12). They are going to fail with quantum algorithms and computing power. Shor's algorithm is the quantum algorithm that threatens the RSA, DSA, and ECDSA cryptographic algorithms.

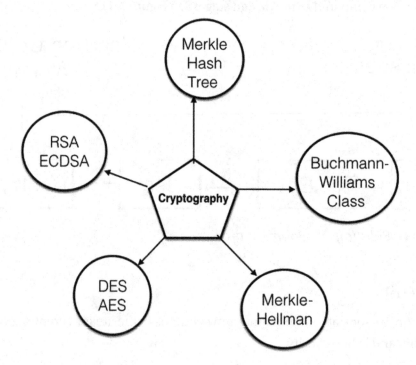

Figure 1-12. *Cryptography models*

A quantum computer–based algorithm can break the symmetric key cryptographic algorithm by a factor of the square root of the size of the key. To find an image of a 256-bit hash function, a quantum algorithm will take only 2^{128} time. Shor's algorithm helps in the quantum factorization of big numbers. This demands that the security systems be quantum-resistant. Elliptic curve cryptography has been shown to improve the cracking process by 21 percent. This method is based on multiplying polynomials and adding random noise.

Quantum algorithms beat the classical equivalents because they can be executed on faster hardware. Quantum algorithms are also based on quantum mechanics. Theoretically, quantum mechanics methods have proven to be faster than classical methods. Quantum mechanics is related to the complex subatomic particle interaction. The subatomic particles can be electrons. Electrons can be present in multiple discrete states at the same time, which is called *superposition.*

Heisenberg's uncertainty principle states that a quantum system has information about an object's velocity and position. Any measurement of velocity will have an impact on the position because the act of observing modifies the state. The subatomic particles can be in an entangled state. A change to one particle impacts another particle when they are separated from each other. Quantum mechanics captures the complexities of the state of electrons using complex numbers.

Grover's algorithm can solve a phone book search on the order of the square root of the number of phone book numbers ($O(\sqrt{n})$ time). Let's say a phone book has 100 million phone numbers. Grover's algorithm can find a number with 10,000 steps. Grover's quantum algorithm uses quantum mechanics–based methods to speed up the search.

Optimization

Optimization problems occur in our daily life in different domains such as finance, banking, healthcare, logistics, and automotive sectors. In wealth management, portfolio selection is based on finding the best assets to balance the risk with the expected gains. A portfolio selection is an NP-hard problem. An NP-hard problem is a difficult and not impossible problem for current computers to solve quickly. Quantum annealing is a quantum algorithm to find solutions for these NP-hard problems.

In the annealing method, the problem is encoded to a similar physical problem. The lowest energy state of a system is the same as the solution to the problem at hand. This system is built in the annealer, and it evolves toward the lowest energy state. In real life, the annealer reaches the lowest energy through the cooling process. In quantum annealing, the tunneling events move the system to minima quickly. See Figure 1-13.

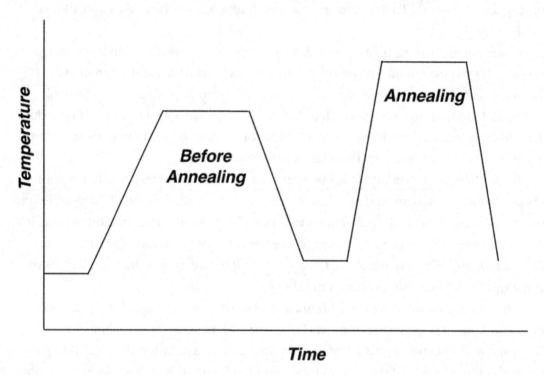

Figure 1-13. *Quantum annealing process*

Quantum annealing techniques help in finding optimal trading paths. A large set of trading orders gets executed, and this impacts the asset price. The execution costs need to be accounted for during the trading execution. Algorithms need to set the trading rate for big orders. This helps in balancing the speed of trading, minimizing the exposure to market risk, and minimizing the impact on the market.

The other area is related to optimizing the arbitrage opportunities. Arbitrage is related to making a profit by taking advantage of the difference in asset prices with less risk. The example given is typically related to currency exchange from euros to dollars, dollars to yens, and yens to euros. This transaction might yield a small profit due to the differences in prices. This is called *cross-currency arbitrage*. This problem is NP-hard to find optimal arbitrage opportunities across the assets in different markets. The quantum annealer method helps in finding the path with the lowest cost.

Credit scoring is another area that can be solved using quantum annealing. Credit scoring of an individual is related to the person's income, age, financial history, collateral, and current liabilities. The risk assessment looks at various factors in the customer's profile. This is a computationally challenging problem, which is an NP-hard problem.

The vehicle routing problem is another optimization problem that can be solved using quantum annealing. Dantzig and Ramser proposed in 1959 the vehicle routing problem. The goal for the optimization is to identify the optimal routes for a set of vehicles serving customer orders. The solution is to identify the set of routes that are part of the order delivery from the source to the destination. The number of vehicles needed for customer orders is also an optimization goal. Quantum annealing is the optimization technique used to determine the local minima route over a given set of candidate functions. This is a method of discretizing a function with many local minima to determine the observables of the function. The quantum annealing process is different from simulated annealing, as it is based on the quantum tunneling process. The particles tunnel through energy barriers from a high state to a low state.

Quantum annealing starts from a superposition of different states of a system that are weighted equally. The time-dependent Schrödinger equation drives the time evolution of the system. This system serves to affect the amplitude of all states as the time increases. Eventually, the ground state is reached to give the instantaneous Hamiltonian of the system.

Cybersecurity

There are new applications in the post-quantum security area that operate on data at rest, in transit, in use, and in motion. These solutions work on data access, business processes, multiparty authentication, and multiparty authorization. These products are based on the algorithms and techniques to tackle man-in-middle detection, quantum security encryption, phishing resistance, biometric security, accountability, and segregation of duties. See Figure 1-14.

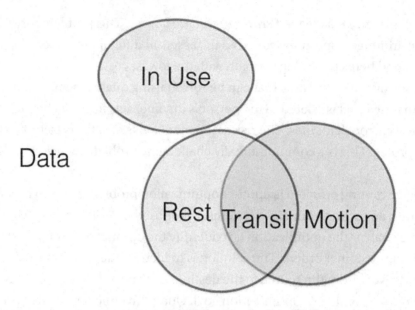

Figure 1-14. *Data in use/rest/transit/motion*

Data at rest can be intercepted and misused by an insider and an external hacker. Data in transit can be modified by the hackers for misuse in the networks. Confidential business data and personally identifiable information are the use cases for data in transit. Mission-critical data cannot be secured by using role-based access and controls. Data governance is needed in an enterprise to protect and secure the data.

Cybersecurity solutions are related to encrypt, decrypt, sign, and verify customer transactions and data. A hacker can intercept and steal the secured data. Secured data has information related to the customer's credit card numbers and Social Security numbers. Using a quantum computer, the hacker has a higher computational power. Quantum algorithms like Shor's are proven to break the current cryptographic systems. The post-quantum cryptographic designers are improving the efficiency of the new encryption and decryption algorithms. Hybrid systems and high-speed quantum-resistant algorithms are required to strengthen post-quantum cryptography algorithms. McEliece public-key encryption, NTRU public-key encryption, and lattice-based public key encryption systems are proven to be quantum-resistant.

Summary

In this chapter, we looked at different solutions in cryptography, optimization, and cybersecurity and how quantum computing is applied and solutions are developed.

## CHAPTER 2

Mathematics Behind Quantum Computing

Introduction

Mathematics is the language in which the gods speak to people.

—Plato

In this chapter, we cover the mathematical concepts related to quantum computing. We look at quantum operators, sets, vectors, matrices, and tensors.

Initial Setup

You need to set up Python 3.5 to run the code samples in this chapter. You can download it from `https://www.python.org/downloads/release/python-350/`.

Quantum Operators

Binary form numbers are used to model gates for a classical computer. Quantum mechanical principles are used for modeling the quantum gates and quantum operations for a quantum computer.

To start, let's look at a Hilbert space (say H). A Hilbert space is a vector space consisting of an inner product of the vectors. These vectors are the unit vectors. They represent the states of the system.

B. Kommadi, *Quantum Computing Solutions*, https://doi.org/10.1007/978-1-4842-6516-1_2

Figure 2-1 defines the inner product of two vectors z and u. The vector is defined on the Hilbert space, and z^* is a complex conjugate of z. An orthonormal basis is defined on a Hilbert space. A two-spin quantum system is represented by states that are two orthonormal basis. An orthonormal set of vectors N in H has elements of space N where each element is a unit vector and any two unique elements are orthogonal.

$$\langle z, u \rangle = \sum_k z^*_k u_k$$

Figure 2-1. *Inner product*

Let's look at the Dirac notation to represent the basis states of a quantum system; see Figure 2-2. The Dirac notation is better than vector-based Heisenberg notation that is used in computer calculations.

$$|\uparrow\rangle = |0\rangle = \begin{bmatrix} 1 \\ 0 \end{bmatrix}$$

$$|\downarrow\rangle = |1\rangle = \begin{bmatrix} 0 \\ 1 \end{bmatrix}$$

Figure 2-2. *Dirac notation*

Now, let's look at Kronecker's product; see Figure 2-3. Kronecker's product is related to the combination of quantum bits. This is used to calculate the product of quantum bits in a multiquantum bit system.

$$|00\rangle = \begin{bmatrix} 1 \\ 0 \end{bmatrix} \otimes \begin{bmatrix} 1 \\ 0 \end{bmatrix} = \begin{bmatrix} 1 \\ 0 \\ 0 \\ 0 \end{bmatrix}$$

$$|01\rangle = \begin{bmatrix} 1 \\ 0 \end{bmatrix} \otimes \begin{bmatrix} 0 \\ 1 \end{bmatrix} = \begin{bmatrix} 0 \\ 1 \\ 0 \\ 0 \end{bmatrix}$$

$$|10\rangle = \begin{bmatrix} 0 \\ 1 \end{bmatrix} \otimes \begin{bmatrix} 1 \\ 0 \end{bmatrix} = \begin{bmatrix} 0 \\ 0 \\ 1 \\ 0 \end{bmatrix}$$

$$|11\rangle = \begin{bmatrix} 0 \\ 1 \end{bmatrix} \otimes \begin{bmatrix} 0 \\ 1 \end{bmatrix} = \begin{bmatrix} 0 \\ 0 \\ 0 \\ 1 \end{bmatrix}$$

Figure 2-3. *Kronecker's product*

Let's look at the quantum gates used for the quantum logic and circuit design. To start, let's talk about digital circuits and traditional digital gates. Digital circuits are classified as sequential and combinational circuits. Sequential circuits are circuits that have the same state of the system, and the input/output of the system is a function of time. Combinational circuits are based on the circuit whose output depends on the input only. MUX, DEMUX, encoder, and decoder are good examples of combinational circuit-based applications.

Memory registers are used for storing and processing the data. The traditional digital gates used for these circuits are mentioned here:

- *AND gate*: AND operation

- *OR gate*: OR operation

- *NOR gate*: Inverse of OR operation

- *NAND gate*: Inverse of AND operation

- *XOR gate*: Exclusive-OR operation

- *XNOR gate*: Inverse of the exclusive-OR operation

- *NOT gate*: INVERSION operation

A classic computer is based on the arithmetic logic unit (ALU) and the controller. The controller controls the instruction flow and the execution of any instruction steps in the cycle. An ALU is based on a digital circuit. The ALU is used to execute mathematical and logic operations. The central processing unit has an ALU and a control unit (CU). See Figure 2-4.

Figure 2-4. *Digital circuit*

A quantum computer will have a quantum processing unit and quantum circuit for processing information. The quantum circuit has a set of quantum *gates*. The quantum gates are used to modify quantum bits. Quantum gates are used to show the evolution of the quantum states of the quantum bits over time. The quantum state of the quantum bit changes at the output of the quantum gate. Quantum Computers follow the principles of preserving norms and reversibility. Preserving the norm specifies that the sum of normal squared probability amplitudes should be equal to the state after the application of the gate. The evolution of the quantum state, which is not measured, is related to reversibility.

Note   A *quantum processing unit* is a quantum chip and equivalent to the classical central processing unit. A quantum chip consists of the QPU, the control unit, and other components. The number of quantum bits in a chip is related to the digital circuit's quantum circuit components.

Quantum operators are based on mathematical functions using the quantum mechanics principles. They operate on quantum states, and the quantum states are represented by vectors and tensors. See Figure 2-5.

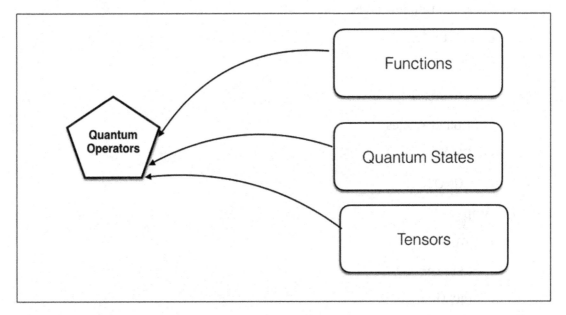

Figure 2-5. *Quantum operators*

Gates are categorized as Hadamard, phase-shifter, and controlled and uncontrolled gates. To give an example, the gates can be X, Y, Z, Rx, Ry, Rz, S, T, CNOT, Toffoli, Swap, and controlled phase-shift gates.

Here is the complete list of quantum gates:

- Pauli gates

  - I, X, Y, Z

- Hadamard gate H

- Phase gates

- PHASE (theta), S, T
- Controlled phase gates
    - CZ
    - CPhase00 (alpha)
    - CPhase01 (alpha)
    - CPhase10 (alpha)
    - Cphase (alpha)
- Cartesian rotation gates
    - RX (theta)
    - RY (theta)
    - RZ (theta)
- Controlled X gates
    - CNOT
    - CCNOT
- Swap gates
    - SWAP
    - CSWAP
    - ISWAP
    - PSWAP (alpha)

A Bloch sphere is used to represent the quantum bit state and the quantum gate transformations of the state. The rotation happens around the x-, y-, and z-axes in three-dimensional space. See Figure 2-6.

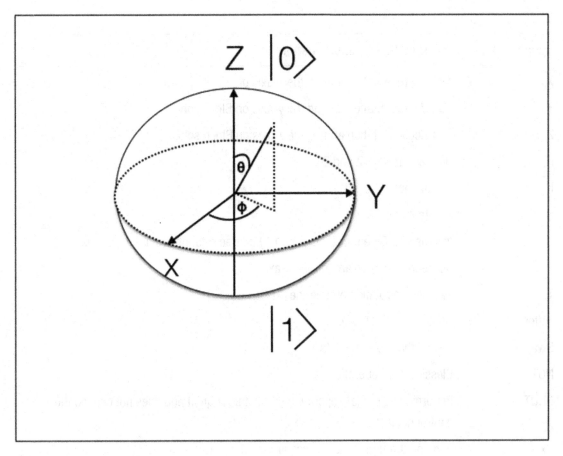

Figure 2-6. *Bloch sphere*

Table 2-1 lists the gate and the quantum state transformation related to each gate over the Bloch sphere.

Table 2-1. *Gate Transformations*

Quantum Gate	Quantum State Transformation
X	180-degree rotation around the x-axis on Bloch sphere
Y	180-degree rotation around the y-axis on Bloch sphere
Z	180-degree rotation around the z-axis on Bloch sphere
H	H gate maps X->Z and Z-X
I	Identity gate
S	S Gate maps X->Y
SI	Inverse of S Gate maps X-> - Y, -180 degree rotation around the z-axis
T	90-degree rotation around the z-axis
TI	-90-degree rotation around the z-axis
Toffoli	Controlled, CNOT gate
Swap	Swaps the state of qubits
NOT	Classical NOT operation
CNOT	Performs Pauli X gate operation on the target qubit and does not change the control qubit
Rx	Rotation through theta degrees around the x-axis
Ry	Rotation through theta degrees around the y-axis
Rz	Rotation through theta degrees around the z-axis
X90	-90-degree rotation around the x-axis
Y90	90-degree rotation around the y-axis
mX90	-90-degree rotation around the x-axis
mY90	-90-degree rotation around the y-axis
CZ	Cphase gate
CRk	Parametrized controlled phase shift with Pi/2k degrees
CR	Controlled phase shift with theta degrees

Now let's look at a code sample in Python to show a quantum operation's Hadamard, Swap, and Identity gates.

Code Sample

quantum_operations.py

```python
import numpy as nump

def multi_quantum_state(*args):
    ret = nump.array([[1.0]])
    for q in args:
        ret = nump.kron(ret, q)
    return ret

zero_quantum_state = nump.array([[1.0], [0.0]])
one_quantum_state = nump.array([[0.0], [1.0]])

Hadmard_quantum_gate = 1.0 / 2**0.5 * nump.array([[1, 1], [1, -1]])

new_quantum_state = nump.dot(Hadmard_quantum_gate, zero_quantum_state)
print("dot product : Hadamard quantum gate and zero  quantumstate",new_
quantum_state)

SWAP_quantum_gate = nump.array([[1,0,0,0],
                    [0,0,1,0],
                    [0,1,0,0],
                    [0,0,0,1]])

t0_quantum_state = multi_quantum_state(zero_quantum_state, one_quantum_state)
t1_quantum_state = nump.dot(SWAP_quantum_gate, t0_quantum_state)
print("SWAP Quantum  Gate - T1 Quantum state",t1_quantum_state)

Identity_quantum_gate = nump.eye(2)
t0_quantum_state = multi_quantum_state(zero_quantum_state, one_quantum_state)
t1_quantum_state = nump.dot(multi_quantum_state(Hadmard_quantum_gate,
Identity_quantum_gate), t0_quantum_state)
print("Multi Quantum State of H Quantum gate and I Quantum gate",t1_quantum_
state)
```

Command for Execution

```
pip3 install numpy
python3 quantum_operations.py
```

Output

```
dot product : Hadamard quantum gate and zero  quantumstate [[0.70710678]
[0.70710678]]
SWAP Quantum  Gate - T1 Quantum state
[[0.] [0.] [1.] [0.]]
Multi Quantum State of H Quantum gate and I Quantum gate
[[0.] [0.70710678] [0.] [0.70710678]]
```

Sets

An object is anything in the universe. It can be living or nonliving. A *set* is a group of objects. The objects in the set are referred to as *members* (or *elements*). The following are examples of a set:

- Consonants in the English language

- Prime numbers

- Arithmetic sequence of numbers

Capital letters are used to represent sets. The elements of the set are represented by lowercase letters. A good example is that N represents the set of natural numbers. Sets are represented in roster form using curly brackets: { }. For example, A = {1,2,3,4,5} is a set of all integers less than 6. The members of the set are specified only once. The order is not important. Another way to represent a set is the set builder form. C = {y : y is a consonant in the English alphabet}.

Sets are classified into the following types:

- Empty set

- Finite set

- Infinite set

- Singleton set

The number of members in an empty set is 0. For a singleton set, it is 1. A finite set will have a finite number of elements. A good example of an empty set is the set of real numbers whose square is -2.

The order of a set is the number of unique members in the finite set. This is also referred to as the *cardinal number* of the set. Two sets are equivalent if their cardinal number is the same. A subset is a set whose elements are members of the superset. For example, A = {1,2} and B = {1,2,3,4}. In this case, A is a *subset* of B, and B is a *superset* of A.

A *power set* is a set of all subsets of a given set. A universal set U is a superset of the given sets. Two sets are disjoint if the intersection of the sets is an empty set. A Venn diagram is used to express the relationships between sets (Figure 2-7).

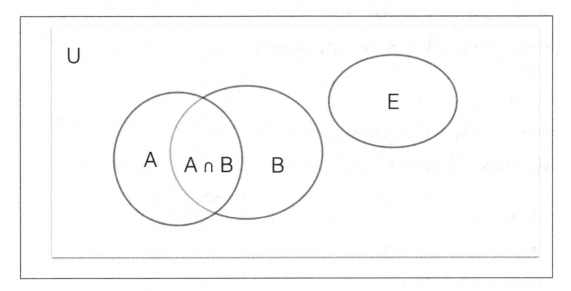

Figure 2-7. *Venn diagram*

A *union* of two sets is a set that has elements in A or B or in both. For example, A = {1,2,3}, B = {4,5,6}, and A U B = {1,2,3,4,5,6}.

The *intersection* of two sets is the set containing all the common elements of the two sets. For example, A = {1,2,3}, B = {2,4,5}, and A ∩ B = {2}.

The *difference* of two sets A and B is the set of elements that belong to A and not B. For example, A = {1,2,3}, B = {2,4,5}, and A - B = {1,3}.

The *complement* of a set is the set that contains the elements of the universal set U and does not have the elements of A.

Now let's look at a code sample in Python to show the set operations such as union, intersection, difference, and symmetric difference.

Code Sample

set operations.py

```python
primes11={3,5,7,11,14};

primes12={3,5,7,11,13,17,19,34}

primes=primes11 & primes12

print("union of sets", primes11,",",primes12," is")
print(primes)

primes1={3,5,7,11,13};

primes2={3,5,7,11,13,17,19,23}

primes=primes1 & primes2

print("intersection of sets", primes1,",", primes2," is")
print(primes)

primes3={3,5,7,11,13,17,19}

primes4={2,3,5,7,11};

primes=primes3-primes4

print("difference of sets",primes3,",",primes4," is")

print(primes)
```

```
primes4={3,5,7,11,13,17,19}

primes5={3,5,7,11,91,101};

primes=primes4 ^ primes5

print("symmetric difference of sets",primes4,",",primes5," is")

print(primes)
```

Command for Execution

```
python3 set_operations.py
```

Output

```
union of sets {3, 5, 7, 11, 14} ,{33, 3, 5, 7, 11, 13, 17, 19} is:
{11, 3, 5, 7}
intersection of sets {3, 5, 7, 11, 14 {34, 3, 5, 7, 11, 13, 17, 19} is:
{11, 3, 5, 7}
difference of sets {3, 5, 7, 11, 13, 17, 19} ,{2, 3, 5, 7, 11} is:
{17, 19, 13}
symmetric difference of sets {3, 5, 7, 11, 13, 17, 19} ,{3, 5, 101, 7, 11,
91} is:
{17, 19, 101, 91, 13}
```

Vectors

A *vector* is defined as an ordered set of n real numbers. Here's an example:

```
C = (c1,c2,c3,.....cn)
```

C is an n-dimensional vector.

Vectors addition is calculated by taking the corresponding dimensional elements and adding them. Similarly, subtraction of the vectors is calculated by subtracting the elements of each dimension.

```
C = (c1,c2,c3,.....cn)
D = (d1, d2, d3,... dn)
C + D = (c1+d1, c2+d2,..., cn+dn)
C - D = (c1-d1, c2-d2,..., cn-dn)
```

Multiplication of vectors by scalars is done by multiplying the scalar to every dimensional element.

```
E = sF = (sf1, sf2, .....,sfn)
```

Now let's look at a code sample in Python to show the vector operations such as dot product, scalar multiplication, sum, difference, product, and division.

Code Sample

vector_operations.py

```
from numpy import array

vector1 = array([1, 2, 3])

vector2 = array([2, 3, 4])
print("dot product of", vector1,", ",vector2,"is")
product = vector1.dot(vector2)
print(product)

scalarval = 0.3
scalarprod = scalarval*vector1
print("scalar multiplied",scalarval,",",vector1,"is")
print(scalarprod)

print("sum of", vector1,", ",vector2,"is")
sum = vector1 + vector2
print(sum)
```

```
print("difference of", vector1,", ",vector2,"is")
diff = vector1 - vector2
print(diff)

print("product of", vector1,", ",vector2,"is")
prod = vector1 * vector2
print(prod)

print("division of", vector1,", ",vector2,"is")
dividedby = vector1 / vector2
print(dividedby)
```

Command for Execution

```
pip3 install bumpy
python3 vector_operations.py
```

Output

```
dot product of [1 2 3] ,  [2 3 4] is
20
scalar multiplied 0.3 , [1 2 3] is
[0.3 0.6 0.9]
sum of [1 2 3] ,  [2 3 4] is
[3 5 7]
difference of [1 2 3] ,  [2 3 4] is
[-1 -1 -1]
product of [1 2 3] ,  [2 3 4] is
[ 2  6 12]
division of [1 2 3] ,  [2 3 4] is
[0.5        0.66666667 0.75       ]
```

Matrices

A *matrix* is the rectangular configuration of the numbers in rows and columns. It is represented by open, (), or closed, [], brackets. The entries in the matrix are called *elements* of the matrix. The order of the matrix is defined as the number of rows and columns. For example, for A = [1 2], 1×2 is the order of the matrix.

A *row* matrix is one that has only one row and any number of columns. A = [1,2,3] is a row matrix. A *column* matrix is the one that has only one column and any number of rows.

Figure 2-8 shows a 2×2 matrix.

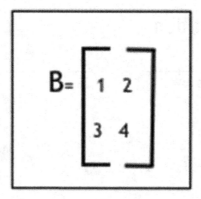

Figure 2-8. *2×2 matrix*

Figure 2-9 shows an example of the column matrix.

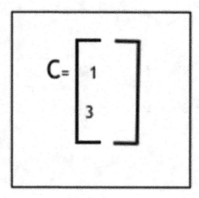

Figure 2-9. *Column matrix*

These are different types of matrices:

- Row matrix

- Column matrix

- Singleton matrix

- Null matrix

- Rectangular matrix

- Square matrix

A *singleton* matrix is a matrix with only one element. For example, A = [1] is a singleton matrix. A *null* matrix is a matrix with all elements zero; for example, A = [0 0] is a zero or null matrix. A *rectangular* matrix is a matrix with the number of rows not equal to the number of columns. See Figure 2-10.

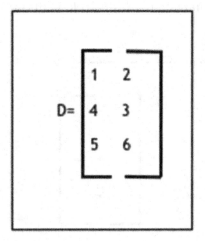

Figure 2-10. *Rectangular matrix*

A *square* matrix is a matrix with the number of rows equal to the number of the columns. See Figure 2-11.

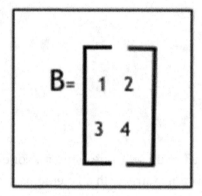

Figure 2-11. *Square matrix*

A *diagonal* matrix is a square matrix with all elements outside the principal diagonal being zeros. See Figure 2-12.

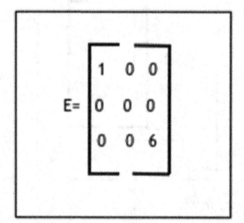

Figure 2-12. *Diagonal matrix*

An *identity* matrix is a square matrix whose elements of the principal diagonal are 1. See Figure 2-13.

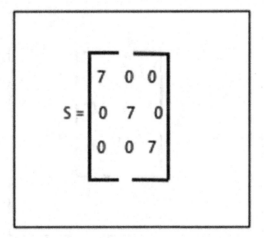

Figure 2-13. *Identity matrix*

A *scalar* matrix is a square matrix whose elements of the principal diagonal are equal, and all nondiagonal elements are zero. See Figure 2-14.

$$S = \begin{bmatrix} 7 & 0 & 0 \\ 0 & 7 & 0 \\ 0 & 0 & 7 \end{bmatrix}$$

Figure 2-14. *Scalar matrix*

A *triangular* matrix is a square matrix whose elements above or below the principal diagonal are equal to zero. The upper triangular matrix has elements below the principal diagonal equal to zero (see Figure 2-15), and the lower triangular matrix has elements above the principal diagonal equal to zero (see Figure 2-16).

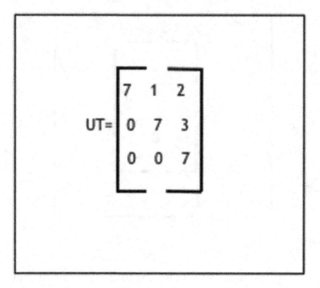

Figure 2-15. *Upper triangular matrix*

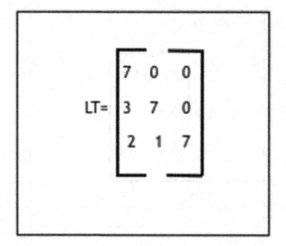

Figure 2-16. *Lower triangular matrix*

The *trace* of a matrix is the sum of diagonal elements of a square matrix. It is represented by tr(). For example, matrix A has a trace equal to tr(A).

Adding matrices is possible for the same types, and the sum is obtained by adding the corresponding elements of the matrices. Similarly, subtraction can be done on the same type, and the result is obtained by subtracting the corresponding elements of the matrices.

Multiplication of matrices is possible if the number of columns in the first matrix is equal to the number of rows in the second one. Scalar multiplication of a matrix is related to the scalar being multiplied to every element of the matrix.

A matrix can be transposed by changing its rows to columns and its columns to rows. The transpose is represented by the superscript [T].

The determinant of a square matrix called matrix A is x1* y2 - x2*y1. See Figure 2-17.

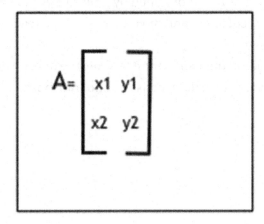

Figure 2-17. *Determinant*

The determinant of the third-order matrix is the sum of the products of the elements of the row or columns with the respective cofactors. Figure 2-18.

x1 X1 +x2 X2 +x3 X3
y1 Y1 + y2 Y2 + y3 Y3
z1 Z1 + z2 Z2 + z3 Z3

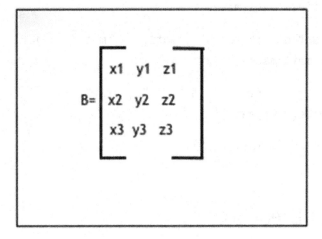

Figure 2-18. *Determinant of third-order matrix*

The inverse of a matrix A is defined as A^{-1}. Therefore, $AA^{-1} = I$ (identity matrix).

```
A⁻¹ = adj A/ det A
```

adj A is the adjoint matrix that is obtained by replacing the elements of a square matrix with the respective cofactors and the transpose of the resultant matrix. det A is the determinant of A.

Now let's look at a code sample in Python to show the matrix operations such as sum, subtraction, product, dot product, division, and transpose.

Code Sample

matrix_operations.py

```python
import numpy

mat1 = numpy.array([[1, 3], [4, 6]])
mat2 = numpy.array([[7, 9], [11, 15]])
print ("Addition of  matrices ", mat1,",", mat2,"is")
print (numpy.add(mat1,mat2))

print ("Subtraction of  matrices ", mat1,",", mat2,"is")
print (numpy.subtract(mat1,mat2))

print ("Product of  matrices ", mat1,",", mat2,"is")
print (numpy.multiply(mat1,mat2))

print ("Dot Product of  matrices ", mat1,",", mat2,"is")
print (numpy.dot(mat1,mat2))

print ("Division of  matrices ", mat1,",", mat2,"is")
print (numpy.divide(mat1,mat2))

print ("Transpose of ", mat1,"is")
print (mat1.T)
```

Command for Execution

```
pip3 install numpy
python3 matrix_operations.py
```

Output

```
Addition of  matrices  [[1 3] [4 6]] , [[ 7  9] [11 15]] is
[[ 8 12]  [15 21]]
Subtraction of  matrices  [[1 3] [4 6]] , [[ 7  9] [11 15]] is
[[-6 -6] [-7 -9]]
Product of  matrices  [[1 3] [4 6]] , [[ 7  9] [11 15]] is
[[ 7 27] [44 90]]
Dot Product of  matrices  [[1 3] [4 6]] , [[ 7  9] [11 15]] is
[[ 40  54] [ 94 126]]
Division of  matrices  [[1 3] [4 6]] , [[ 7  9] [11 15]] is
[[0.14285714 0.33333333] 0.36363636 0.4       ]]
Transpose of  [[1 3] [4 6]] is [[1 4] [3 6]]
```

Tensors

A *tensor* is a multidimensional array (see Figure 2-19). A one-dimensional tensor is a vector. A two-dimensional tensor is a matrix.

```
     v111 v121 v131   v112 v122 v132   v113 v123 v133

T1 = v211 v221 v231   v212 v222 v232   v213 v223 v233

     v311 v321 v331   v312 v322 v332   v313 v323 v333
```

Figure 2-19. *Tensor example*

Addition, subtraction, and multiplication operations can be performed on tensors just like on vectors and matrices.

Now let's look at a code sample in Python to show the set operations such as sum, difference, product, division, and dot product.

Code Sample

tensor_operations.py

```python
from numpy import array
from numpy import tensordot
tensor1 = array([
  [[1,2,3],    [4,5,6],    [7,8,9]],
  [[13,12,13], [14,15,16], [17,18,19]],
  [[24,22,23], [24,25,26], [27,28,29]],
  ])
tensor2 = array([
  [[1,4,3],    [4,5,6],    [7,8,9]],
  [[11,13,13], [14,15,16], [17,18,19]],
  [[21,25,23], [24,26,26], [27,28,29]],
  ])

print("tensor1",tensor1)
print("tensor2",tensor2)
print("sum of tensors", "is")
sum = tensor1 + tensor2
print(sum)

print("difference of tensors",  "is")

diff = tensor1 - tensor2
print(diff)

print("product of tensors",  "is")

prod = tensor1 * tensor2
print(prod)

print("division of tensors",  "is")

division = tensor1 / tensor2
print(division)
```

```
print("dot product of tensors",  "is")

dotproduct = tensordot(tensor1, tensor2, axes=0)
print(dotproduct)
```

Command for Execution

```
pip3 install numpy
python3 tensor_operations.py
```

Output

```
tensor1  [
  [[1,2,3],     [4,5,6],     [7,8,9]],
  [[13,12,13], [14,15,16], [17,18,19]],
  [[24,22,23], [24,25,26], [27,28,29]],
  ]
tensor2 [
  [[1,4,3],     [4,5,6],     [7,8,9]],
  [[11,13,13], [14,15,16], [17,18,19]],
  [[21,25,23], [24,26,26], [27,28,29]],
  ]
sum of tensors is [
[[ 2   6   6] [ 8 10 12] [14 16 18]]
[[24 25 26] [28 30 32] [34 36 38]]
[[45 47 46] [48 51 52] [54 56 58]]
]
difference of tensors is [
[[ 0 -2  0] [ 0  0  0] [ 0  0  0]]
[[ 2 -1  0] [ 0  0  0] [ 0  0  0]]
[[ 3 -3  0] [ 0 -1  0] [ 0  0  0]]
]
product of tensors is [
[[  1   8   9] [ 16  25  36] [ 49  64  81]]
[[143 156 169][196 225 256][289 324 361]]
[[504 550 529] [576 650 676] [729 784 841]]
]
```

```
division of tensors is [
[[1.     0.5    1.    ] [1.    1.     1.    ][1.          1.          1.    ]]
 [[1.18181818 0.92307692 1. ][1.    1.    1. ][1.  1.    1. ]]
 [[1.14285714 0.88        1. ][1.    0.96153846 1.][1.  1.          1. ]]
]
dot product of tensors is
```

Note that the dot product is not shown in the output because of the size of the tensor.

Summary

In this chapter, we looked at quantum operators, sets, vectors, matrices, and tensors. These are the mathematical concepts required for quantum computing.

PART II

Quantum Computing Basics

## CHAPTER 3

Quantum Subsystems and Properties

Introduction

The mathematics of quantum mechanics very accurately describes how our universe works.

—Antony Garrett Lisi

This chapter covers quantum subsystems and their properties. Single and multiple quantum bit systems are discussed in detail. Quantum states, protocols, operations, and transformations are also presented in this chapter.

A quantum computer processes information related to real-life problems and solves them using quantum mechanics. A classical computer uses bits that are either 0 or 1. We have to pick one of those options since it is based on the electric current in a classical computer wire. Quantum units are based on the quantum superposition principle in which 0 and 1 are superposed. This opens up possible states that are infinite and based on the 0 and 1 with weights in a continuous range. In classical computers, decimal system–based values are converted to binary numbers. Similarly, quantum units are defined using the two-dimensional Hilbert space. See Figure 3-1.

© Bhagvan Kommadi 2020
B. Kommadi, *Quantum Computing Solutions*, https://doi.org/10.1007/978-1-4842-6516-1_3

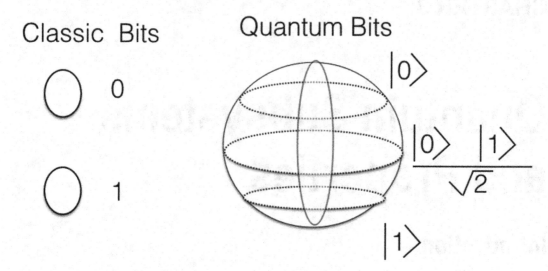

Figure 3-1. *Classic versus quantum bits*

Initial Setup

You need to set up Python 3.5 to run the code samples in this chapter. You can download it from `https://www.python.org/downloads/release/python-350/`.

Single Qubit System

The quantum bit is the foundation for a quantum computation framework. By using quantum bits (*qubits*), the quantum computer performs better than a classical computer. A quantum computer stores the information in quantum bits. A quantum bit represents a quantum particle like an electron having a charge as 0 or 1 and as 0 and 1 simultaneously. A Bloch sphere is used to show the position of the quantum bit's state. The sphere has a radius of one unit. A Hilbert space is used to model the surface of the sphere. A Hilbert space is related to an inner product of vectors in vector space. Length and angle measurements are shown in the inner product. See Figure 3-2.

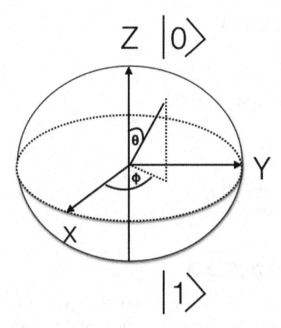

Figure 3-2. *Bloch sphere*

Now let's look at a code sample in Python that demonstrates a single qubit system.

Code Sample

<u>**single_qubit_system.py**</u>

```python
import numpy as nump
import scipy as scip
import scipy.linalg

Zero_state = nump.array([[1.0], [0.0]])
One_state = nump.array([[0.0], [1.0]])

NormalizeQuantumState = lambda quantum_state: quantum_state /
scip.linalg.norm(quantum_state)

Plus_State = NormalizeQuantumState(Zero_state + One_state)

print("normalized quantum state of",Zero_state, "+", One_state,"is",
Plus_State)
```

Command for Execution

```
pip3 install numpy
pip3 install scipy
python3 single_qubit_system.py
```

Output

```
normalized quantum state of [[1.][0.]] + [[0.][1.]] is
[[0.70710678][0.70710678]]
```

Multiple Qubit System

A multiqubit system can be shown using the Kronecker product. Let's look at two quantum particles, c and d. Their state can be modeled as shown here:

```
|ψc>= α0 |0> + βc |1> (Particle C)
|ψdi>= αd|0> + βd |1> (Particle D)
```

A multiple quantum bit register has the quantum system in which the quantum bits are in superposition states. A classical register with s quantum bits has 2s unique states. The quantum system will consist of 2p states in superposition. The system can be in one, two, three, ... all of the 2s states. Let's look at a quantum system with 2 qubits and their possible states (see Figure 3-3).

Qubit1	Qubit2
0	0
0	1
1	0
1	1

Figure 3-3. *Two qubit system*

Now, let's look at a 3 qubit system and the possible states (see Figure 3-4).

Qubit1	Qubit2	Qubit3
0	0	0
0	0	1
0	1	0
0	1	1
1	0	0
1	0	1
1	1	0
1	1	1

Figure 3-4. *Three qubit system*

The quantum states of the quantum bits in a multiple-qubit system will be a vector with complex coefficients. The real logical state is referred to as *observable*. The state is measured, and the result is observed. A quantum system will have filters, which are used to identify the quantum state. Different functions are modeled based on the different sets of observables.

A 2 qubit system is specified using the following equation:

$$q(0) + q(1) = 1$$

Single quantum bit measurements are used to model 2 qubit measurement operations. A quantum register can be measured partially. The unmeasured leftover is transformed using the identity quantum operation.

Quantum States

A quantum bit exists in 0, 1, or a state between 0 and 1 quantum states. This helps a quantum computer to model a huge set of possibilities through the quantum state. S bits are equivalent to 2^S multiple states. The state between 0 and 1 is referred to as *quantum superposition*. Superposition is modeled as a wave function. A wave function can have infinite states mathematically.

A quantum state is the state of the isolated quantum system. It defines the probability distribution of the measured observable quantum bit. Quantum states are classified as follows:

- Pure

- Mixed

Two quantum states can be mixed with one another to create another quantum state. This is similar to the mixing of matter waves in physics.

A Bloch sphere is used to present the quantum states (see Figure 3-5). A quantum bit can be in a superposition state, as shown here:

```
amp1|0> + amp2|1> = |amp>  where |0> and |1> are column vectors
```

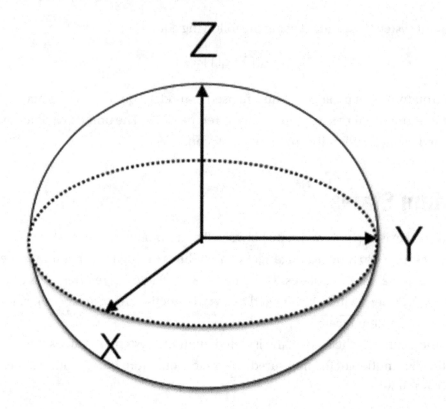

Figure 3-5. *Bloch sphere*

amp1 and amp2 are called *probability amplitudes*, which are complex numbers.

The norm squares and probabilities are calculated as a result of the measurement of a quantum bit.

```
|amp1^2| + |amp2^2| = 1
```

A superposition state can also be represented as shown here (see Figure 3-6):

$$|\psi\rangle \;=\; 1/\sqrt{2} * (|0\rangle + |1\rangle)$$

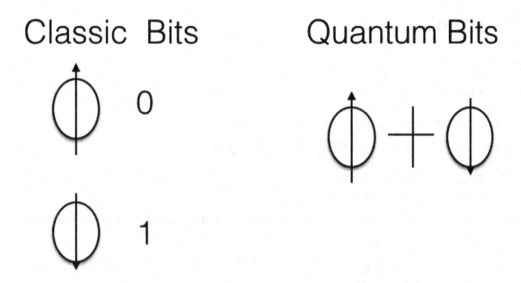

Figure 3-6. *Superposition*

Figure 3-6 shows the quantum bits in superposition compared to the classical bits' 0 and 1 states.

A quantum particle's state is modeled as a quantum bit. Figure 3-7 shows the superposition state of the quantum particle.

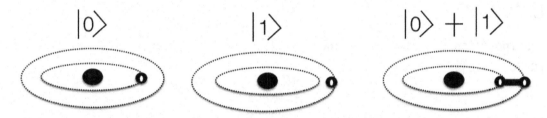

Figure 3-7. *Superposition state of quantum particle*

Now let's look at a code sample in Python to demonstrate the superposition of the quantum bits and states.

Code Sample

<u>superposition.py</u>

```python
import numpy as nump

zero_quantum_state = nump.array([[1.0], [0.0]])
one_quantum_state = nump.array([[0.0], [1.0]])

quantum_amp1 = 1.0 / 2**0.5
quantum_amp2 = 1.0 / 2**0.5
superposition_quantum_state = quantum_amp1 * zero_quantum_state + quantum_
amp2 * one_quantum_state
print("superposition of", zero_quantum_state, " and ", one_quantum_state,"
is ",superposition_quantum_state)
```

Command for Execution

```
pip3 install numpy
python3 superposition.py
```

Output

```
superposition of [[1.][0.]]  and  [[0.][1.]]  is  [[0.70710678]
[0.70710678]]
```

Quantum Protocols

In this section, we look at entanglement and teleportation, which are based on quantum protocols.

Quantum Entanglement

Quantum entanglement is the interaction of multiple quantum particles with each other. The interaction leads to entanglement, and the quantum particles might separate eventually in the space.

This quantum protocol–based phenomenon was observed in 1935 by Albert Einstein, Boris Podolsky, and Nathen Rosen during a quantum mechanical experiment. They wanted to show the logical impossibility of quantum mechanics.

A change in the quantum state of the entangled system–based particle impacts the other and is detectable. The particles in an entangled system are aware of each other. See Figure 3-8.

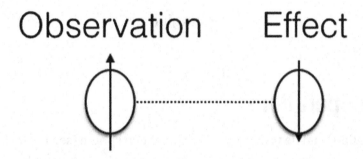

Figure 3-8. *Entanglement*

Quantum Teleportation

Quantum teleportation is the transfer of information from a source to a destination. Classical communication networks and channels are used to transfer information. The teleportation based on quantum protocols happens when there is an entanglement of quantum particles at the source and the destination. The current record is a distance of 1,400 kilometers (about 870 miles). Quantum channels can be created between the source and the destination. The quantum state of the quantum particles is teleported from the source to the destination. See Figure 3-9.

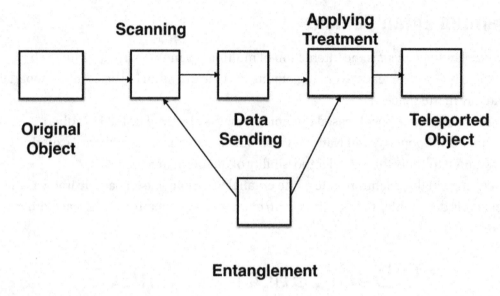

Figure 3-9. *Teleportation*

Quantum Operations

Quantum operations can be visualized using the Bloch sphere (see Figure 3-10). The quantum state of a quantum bit can be shown using a point on the surface of the sphere. The two dials of the sphere indicate the angular and azimuthal coordinates of the sphere. The sphere's surface can be modeled using a Hilbert space. As mentioned, a Hilbert space is based on a vector space of an inner product of two vectors. The inner product helps in the measurement of the length and angle.

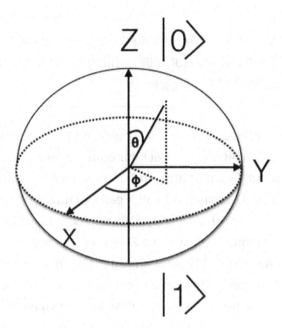

Figure 3-10. *Quantum operations: Bloch sphere*

Note   A Bloch sphere is used to represent the quantum operation. The rotation happens around the x-, y-, and z-axes of a unit radius three-dimensional sphere.

Quantum gates are quantum operations using quantum bits. Quantum operations are the building blocks of the quantum circuits. Quantum circuits are the fundamental blocks on which a quantum computer is built.

Let's start looking at the Hadamard gate first. Hadamard gate is the gate that operates on the base state from |0> to (|0>+|1>)* $1/\sqrt{2}$ and |1> to (|0> - |1>)* $1/\sqrt{2}$. This gate transformation results in a superposition.

A Hadamard gate is expressed as a matrix, as shown here:

H = $1/\sqrt{2}$ (|0> + |1>) <0| + $1/\sqrt{2}$ (|0> - |1>) <1|

Now, we look at Swap gates. A Swap gate swaps the quantum state of the gate.

It can be represented as follows in matrix form:

SWAP = |01>

> **Note**   Dirac notation is used for writing equations. Paul Dirac was the first person to come up with a unique notation for representing vectors and tensors. This notation is efficient compared to matrices.

An X gate is a quantum equivalent of a NOT gate. A CNOT gate is a controlled-NOT gate. A CNOT gate negates a bit; the control bit is equal to only 1. A Toffoli gate is a controlled-NOT gate, which is a quantum equivalent of an AND gate. Toffoli negates a bit if both control bits are equal to only 1. A Pauli X gate is a quantum operation that rotates a quantum bit over the x-axis with an angle of 180 degrees. A Pauli Z gate is a quantum operation that rotates a quantum bit over the z-axis with an angle of 180 degrees.

Now let's look at some CNOT, X, Z, and Measure gate implementations to demonstrate the quantum gates. The implementations of a CNOT gate, X gate, Z gate, normalize operation, and Measure gates in Python are shown next.

Code Sample

Multi_Qubit_System.py

```python
import math
import random

def ApplyCNOTQuantumGate(zerozero,zeroone,onezero,oneone):
    onezero, oneone = oneone, onezero
    return GetQuantumbits(zerozero,zeroone,onezero,oneone)

def ApplyHQuantumGate(zerozero,zeroone,onezero,oneone):
    a = zerozero
    b = zeroone
    c = onezero
    d = oneone

    zerozero = a + c
    zeroone  = b + d
    onezero  = a - c
    oneone   = b - d
```

```python
        normalize()

        return GetQuantumbits(zerozero,zeroone,onezero,oneone)

def ApplyXQuantumGate(zerozero,zeroone,onezero,oneone):
        a = zerozero
        b = zeroone
        c = onezero
        d = oneone

        zerozero = c
        zeroone  = d
        onezero  = a
        oneone   = b

        return GetQuantumbits(zerozero,zeroone,onezero,oneone)

def ApplyZQuantumGate(zerozero,zeroone,onezero,oneone):
        onezero *= -1
        oneone  *= -1

        return GetQuantumbits(zerozero,zeroone,onezero,oneone)

def ApplyNormalization(zerozero,zeroone,onezero,oneone):
        norm = (abs(zerozero) ** 2 + abs(zeroone) ** 2 +
                abs(onezero) ** 2 + abs(oneone) ** 2) ** 0.5
        zerozero /= norm
        zeroone  /= norm
        onezero  /= norm
        oneone   /= norm
        return GetQuantumbits(zerozero,zeroone,onezero,oneone)

def MeasureQuantumbit(zerozero,zeroone,onezero,oneone):
        zerozeroprob = abs(zerozero) ** 2
        zerooneprob  = abs(zeroone)  ** 2
        onezeroprob  = abs(onezero)  ** 2
        randomchoice = random.random()
```

```
    if randomchoice < zerozeroprob:
        zerozero = complex(1)
        zeroone  = complex(0)
        onezero  = complex(0)
        oneone   = complex(0)
        return (0, 0)
    elif randomchoice < zerooneprob:
        zerozero = complex(0)
        zeroone  = complex(1)
        onezero  = complex(0)
        oneone   = complex(0)
        return (0, 1)
    elif randomchoice < onezeroprob:
        zerozero = complex(0)
        zeroone  = complex(0)
        onezero  = complex(1)
        oneone   = complex(0)
        return (1, 0)
    else:
        zerozero = complex(0)
        zeroone  = complex(0)
        onezero  = complex(0)
        oneone   = complex(1)
        return (1, 1)

def GetQuantumbits(zerozero,zeroone,onezero,oneone):
    comp = [zerozero, zeroone, onezero, oneone]
    comp = [i.real if i.real == i else i for i in comp]
    comp = [str(i) for i in comp]
    comp = ["" if i == "1.0" else i for i in comp]

    ls = []
    if abs(zerozero) > 0:
        ls += [comp[0] + " |00>"]
    if abs(zeroone)  > 0:
        ls += [comp[1] + " |01>"]
```

```
    if abs(onezero)  > 0:
        ls += [comp[2]  + " |10>"]
    if abs(oneone)   > 0:
        ls += [comp[3]   + " |11>"]

    comp = " + ".join(ls)

    return comp

a_quantum_state = 1
b_quantum_state = 0
c_quantum_state = 0
d_quantum_state = 0
zerozero_quantum_state = complex(a_quantum_state)
zeroone_quantum_state  = complex(b_quantum_state)
onezero_quantum_state  = complex(c_quantum_state)
oneone_quantum_state   = complex(d_quantum_state)
result1 = ApplyCNOTQuantumGate(zerozero_quantum_state,zeroone_quantum_
state,onezero_quantum_state,oneone_quantum_state)
print("cnot gate operation - result",result1)
result2 = ApplyXQuantumGate(zerozero_quantum_state,zeroone_quantum_
state,onezero_quantum_state,oneone_quantum_state)
print("xgate operation - result",result2)
result3 = ApplyZQuantumGate(zerozero_quantum_state,zeroone_quantum_
state,onezero_quantum_state,oneone_quantum_state)
print("zgate operation - result",result3)
result4 = ApplyNormalization(zerozero_quantum_state,zeroone_quantum_
state,onezero_quantum_state,oneone_quantum_state)
print("normalize operation - result",result4)
result5 = MeasureQuantumbit(zerozero_quantum_state,zeroone_quantum_
state,onezero_quantum_state,oneone_quantum_state)
print("measure gate operation - result",result5)
```

Command for Execution

```
python3 Multi_Qubit_System.py
```

Output

```
cnot gate operation - result  |00>
xgate operation - result  |10>
zgate operation - result  |00>
normalize operation - result  |00>
measure gate operation - result (0, 0)
```

Quantum Transformations

Quantum transformations are related to the quantum state transformations using quantum bits. We look at the Kronecker transformation and measure gates in this section.

Kronecker Transformation

A tensor product is used to model a multiquantum bit system. A Kronecker product is a tensor product of vectors and matrices.

A multiquantum state is represented using multiple quantum bits and vectors. The dimensionality of the quantum state increases exponentially with the number of quantum bits in the system.

Note   The tensor product between two quantum states is represented using the symbol $\otimes$. The multiquantum bit state is shown using Dirac notation as |100>, which is the same as |1> $\otimes$ |0> $\otimes$ |0>.

Measure Gate Transformation

Now we look at the Measure gate. A Measure gate transforms the information using quantum operations. Measurement transformation can be represented using the following equation:

```
amp1|0> + amp2|1> = |amp>
```

where amp1 and amp2 are amplitudes of the quantum bits.

You need to define the projectors to simulate measurement transformation.

Proj0 = |0> <0|
Proj1 = |1> <1|

The state defined here shows the measure gate transformation:

|ψ> = (1/4) (|00> + |01> + |10> +|11>)

Now let's look at a code sample in Python to demonstrate a Kronecker transformation in Python.

Code Sample

multi_kronecker.py

```
import numpy as nump

def multi_quantum_state(*args):
    ret = nump.array([[1.0]])
    for q in args:
        ret = nump.kron(ret, q)
    return ret

zero_quantum_state = nump.array([[1.0], [0.0]])
one_quantum_state = nump.array([[0.0], [1.0]])

three_quantum_states = nump.kron(nump.kron(zero_quantum_state, one_quantum_
state), one_quantum_state)
print("three  quantum states -kronecker",three_quantum_states)
```

Command for Execution

```
pip3 install numpy
python3 multi_kron.py
```

Output

```
three   quantum states -kronecker product
[[0.][0.]
 [0.][1.]
 [0.][0.]
 [0.][0.]]
```

Summary

In this chapter, we looked at single and multiple quantum bit systems. Quantum bits system phenomena such as entanglement, superposition, and teleportation were discussed in detail. Finally, the quantum states, gates, and transformations were presented.

Quantum Information Processing Framework

Introduction

"A classical computation is like a solo voice—one line of pure tones succeeding each other. A quantum computation is like a symphony—many lines of tones interfering with one another."

—Seth Lloyd

This chapter gives an overview of a quantum information processing framework. You will see how a quantum information processing framework is applied in real life. Code examples are presented for the quantum algorithms discussed. In addition, this chapter discusses quantum circuits in detail.

Initial Setup

You need to set up Python 3.5 to run the code samples in this chapter. You can download it from `https://www.python.org/downloads/release/python-350/`.

Quantum Circuits

A quantum circuit is a key part of quantum computing. As we have seen, quantum computing is based on quantum mechanics principles. For example, quantum mechanics principles such as superposition and quantum entanglement are widely used in quantum computing. Quantum bits are the same as the classical bits. The Hilbert state space has the eigenvalues related to the linear superposition. The state values will be |0> and |1>. These eigen states are operated on using unitary operations.

© Bhagvan Kommadi 2020
B. Kommadi, *Quantum Computing Solutions*, https://doi.org/10.1007/978-1-4842-6516-1_4

Let's look at what a Hilbert space is. A Hilbert space is similar to an inner product space of finite-dimensional complex vectors. The inner product of two complex vectors is the same as the inner product of two matrices that represent the vectors. Now let's see what eigenvalues are. *Eigenvalues* are complex numbers that satisfy the linear equation B |v > = v|v>. |v> is a nonzero vector, which is the eigenvector with linear operator B.

Let's look at the superposition principle that was discussed in earlier chapters. The superposition principle in quantum mechanics is related to the states of the quantum system. Let's say |u> and |v> are the two states of the quantum system. Superposition α|u> + β |v> is a state that is valid if $|\alpha|^2 + |\beta|^2 = 1$. Schrodinger's equation is another principle from quantum mechanics that is used in quantum computing. The quantum system's state evolves with time. The time evolution is explained mathematically by Schrodinger's equation.

Figure 4-1 shows a sample quantum circuit with Hadamard gates. The input and output will be in quantum bits. The Hadamard gate was introduced in earlier chapters. It is a popular quantum gate in terms of usage. It operates on the single quantum bit gates by rotation and reflection on the Bloch sphere. The rotation of 90 degrees is around the y-axis, and the x-axis rotation is 180 degrees.

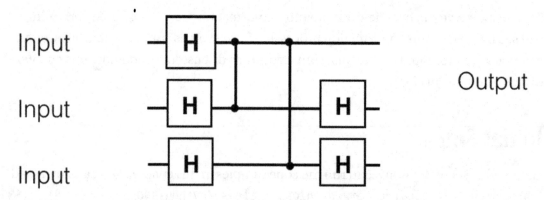

Figure 4-1. Quantum circuit

The quantum circuit model is based on the quantum entanglement of quantum bits. Quantum entanglement has no classical equivalent. The superposition principle is used in quantum computers to save data in exponential size compared to classical computers. Classical computers using M bits can save one out of the 2^M combinations. Quantum computers can save 2^M combinations of information. Let's see what quantum entanglement is. We discussed quantum entanglement in the initial chapters.

Entanglement is related to a pair of entangled particles whose state is dependent on each other. This phenomenon is observed when they are separated by huge distances.

Quantum circuits are used to create a quantum computer. The circuit consists of wires and quantum gates. Quantum logic gates are based on either a quantum bit or multiple quantum bit gates. These gates are reversible. Quantum gates are used to transmit and modify the information.

Now let's look at the controlled-NOT gate circuit shown in Figure 4-2. A controlled-NOT gate is a multiquantum bit logic gate. It is one of the universal quantum bit gates. A CNOT gate has two input quantum bits. These quantum bits are the control quantum bit and the target quantum bit. Similarly, the CNOT operator takes 2 qubits as input. For the pair of input quantum bits, one is called a *controlled* quantum bit and the other the *target* quantum bit. When you change the controlled quantum bit to |1>, the target quantum bit's value is flipped. Otherwise, the target quantum bit remains the same.

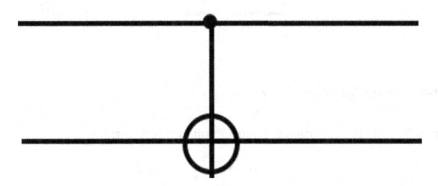

Figure 4-2. *CNOT gate circuit*

Figure 4-3 shows a CNOT gate circuit in matrix form.

$$CNOT = \begin{bmatrix} 1 & 0 & 0 & 0 \\ 0 & 1 & 0 & 0 \\ 0 & 0 & 0 & 1 \\ 0 & 0 & 1 & 0 \end{bmatrix}$$

Figure 4-3. *CNOT gate circuit in matrix form*

Let's look at another type of circuit. A controlled-U gate circuit is based on a controlled-U gate. This gate operates on the input of 2 qubits. The first quantum bit is a control quantum bit. Figure 4-4 shows the circuit. The controlled-U gate has U, which is a unitary matrix. The unitary matrix operates on a number m of quantum bits. The basic example of a controlled-U gate is a controlled-NOT gate.

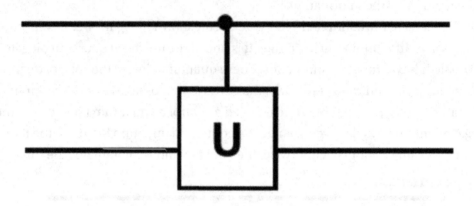

Figure 4-4. *Controlled-U gate circuit*

Figure 4-5 shows the controlled-U gate circuit in matrix form.

$$CU = \begin{bmatrix} 1 & 0 & 0 & 0 \\ 0 & 1 & 0 & 0 \\ 0 & 0 & U_{00} & U_{01} \\ 0 & 0 & U_{10} & U_{11} \end{bmatrix}$$

Figure 4-5. *Controlled-U gate circuit matrix*

A Toffoli gate circuit is another type of circuit, as shown in Figure 4-6. A Toffoli gate combined with a Hadamard gate is a 3 qubit logic gate.

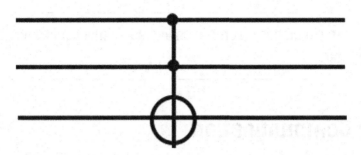

Figure 4-6. *Toffoli gate circuit*

A Toffoli gate has an input of 3 bits and an output of 3 bits. Out of the three input bits, two of them are control bits, and the third bit is a target bit. If control bits have a value of 1, the target bit is flipped. If the control bits do not have a value of 1, the control bits are not changed.

Figure 4-7 shows a Toffoli gate circuit in matrix form.

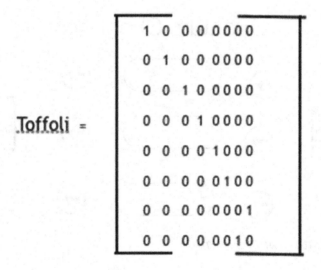

Figure 4-7. *Toffoli gate circuit matrix*

After looking at the different types of quantum circuits, the last step in the quantum circuit model is the measurement projected on the computational scale. This step is not reversible. There are two types of measurement: deferred and implicit. The *deferred* measurement method allows changing the quantum operations to the classical equivalents. In this method, a measurement can be sent from the middle stage to the end of the circuit. In the *implicit* measurement method, quantum bits that are not measured can be measured implicitly. In this method, generality loss does not happen.

Note   A quantum circuit has reversible changes to a mechanical analog, which is a quantum register.

Quantum Communication

Quantum communication is the process of sending the quantum state from one point to another, as shown in Figure 4-8. Superdense coding is a popular communication protocol that was first proposed in 1992 by Bennett and Wiesner. *Quantum teleportation* is related to the movement of quantum states without quantum communication. This is an inverse method to superdense coding. Quantum teleportation is a method for shifting the quantum states without the quantum communication channel between the receiver and the sender.

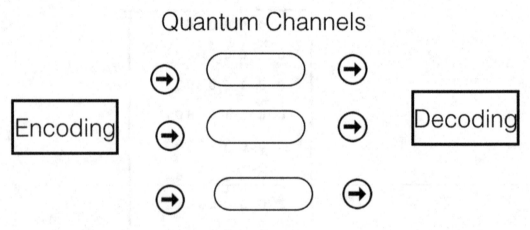

Figure 4-8. *Quantum communication*

In superdense coding, classical information bits are transferred from the source to the destination. In quantum teleportation, quantum bits are transferred by using the communication channel between two classical bits.

The superdense protocol has an instruction set and the output results. The protocol is valid if it generates the expected output. The superdense protocol is based on quantum entanglement. Quantum entanglement occurs when a single common source can produce multiple quantum systems. The state of the quantum systems does not follow Bell inequality rules. The systems that are entangled impact each other even if they are separated by huge distances.

We have looked at the superdense quantum communication protocol; the other protocols are the continuous variable, satellite, and linear optics quantum communication protocols. The electric field of the incident light is measured using a homodyne detector. The measurement results in a continuous value. Continuous variable quantum communication is based on the homodyne detector measurement outcome. The satellite communication protocol is based on ground-to-satellite and satellite-to-satellite communication channels. The linear optical quantum communication protocol is based on optical fiber–based communication channels.

Quantum Noise

Quantum noise occurs because of uncertainty of a quantum origin–based physical entity. Spectral density is used to measure the noise intensity. This is measured at a specific frequency in a time-dependent manner. Quantum devices create noise when the field is amplified. Quantum computers also create noise because of the noncommutation of the modes of the quantum field. Figure 4-9 represents the noise in quantum computing form and the wave form.

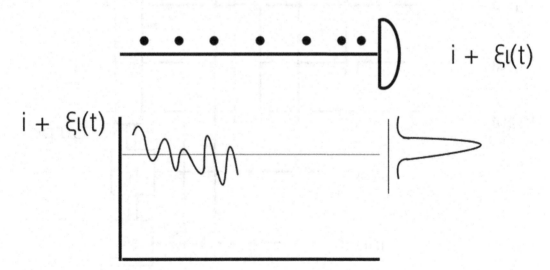

Figure 4-9. *Quantum noise*

H.B.Callen and T.A.Welton represented the spectral density in mathematical equations. Roger H.Koch measured the spectral entities of the quantum noise experimentally first. Spectral density is related to the signal periodograms of time segments. Mathematically, Fourier analysis is a good method to measure spectral density by breaking down the function to a set of periodograms. A periodogram is a time period series in mathematics.

Note   Quantum noise happens because of signal changes in an electric circuit. The signal changes occur because of the electron's discrete character.

Quantum Error Correction

The quantum error correction method helps to shield the quantum information from quantum noise and decoherence errors. The quantum error correction method was first created in 1995. Figure 4-10 shows error correction in the quantum circuit.

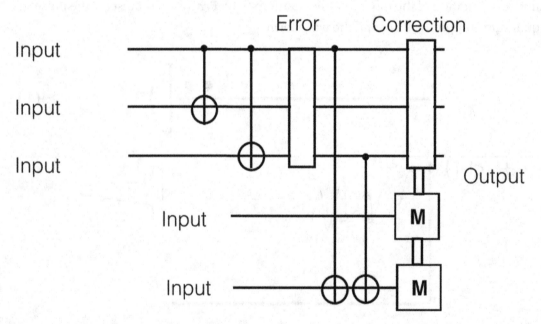

Figure 4-10. *Quantum error correction*

Let's look at classical error correction techniques such as coding. Coding is related to data cloning. This technique is not helpful for quantum error correction because of the no-cloning principle. The no-cloning principle states that a quantum state that is not known cannot be copied exactly. Quantum error correction does not allow a huge number of copies. Errors cannot be prevented by using direct measurement because quantum superposition fails. Quantum errors can happen due to quantum bit flip, phase shift, and continuous errors.

Now let's look at how to fix quantum errors. Coherent techniques help to bring down the control errors. The errors due to decoherence are time dependent. Modeling the errors helps to measure the errors, and positive operator–valued measures are used to measure the errors. Tracing the lost quantum bit is another way to model the quantum bit loss and removal. Correction is done by validating the presence of the quantum bit in the system first. Quantum bits are lost due to controls that are significant. The Shor code protocol designed by Peter Shor helps in identifying and fixing the errors from a quantum bit.

Let's now look at the Shor code protocol. The Shor code is a group of three quantum bit phase flip and bit flip codes. This quantum code helps to guard against the error impact on a quantum bit. Quantum states are operated by the unitary operation to result in a quantum error correction–based code. This result will be part of the subspace D of a bigger Hilbert space. A Hilbert space is an inner product of finite-dimensional complex vectors.

Limitations of Quantum Computing

Let's look at the limitations of quantum computing before we look at different quantum algorithms. Quantum computers are prone to errors and problems due to the environment. Decoherence happens because of the collapse of the state function due to the measurement of state for a quantum bit. Decoherence causes the errors in the computation. The quantum bits count impacts the decoherence probability exponentially. Decoherence occurs because of the quantum bit flip, spontaneous symmetric breaking, and phase flip coding protocols. Quantum bit flip code flips the quantum bit based on a probability. The symmetric breaking protocol is based on a symmetry group F if an F transformation on the quantum system can be achieved by operating transformations on the quantum subsystems. Quantum bit phase flip code is based on the flip of the phase based on a probability.

Small errors can be removed by using error correction protocols in classical computation. In quantum computation, small errors cannot be fixed. Quantum systems are entangled to the environment, and hence the error is induced before the computation begins. Initialization of the quantum bit induces the errors due to unitary transformations.

Quantum Algorithms

Algorithms are used in computation and data processing. The real-life situations are modeled and solved using heuristics and methods. An algorithm consists of a set of commands to execute a task. Quantum algorithms help in providing solutions to complex problems that cannot be solved by classical computers.

Quantum programming languages help to create quantum software with quantum algorithms. A quantum algorithm consists of steps such as encoding the quantum data to quantum bits, a unitary quantum gates chain operating on the quantum bits, and terminating after measuring the quantum bits. A unitary quantum gate is similar to a unitary matrix that operates on the quantum bits. A unitary matrix inverse will be equal to its conjugate transpose.

Quantum computing is based on quantum mechanical properties such as superposition and entanglement. Specifically, quantum algorithms use these principles.

Note   The word *algorithm* word originates from the word *algebra*. It is named after Arabian mathematician Al Khwarizmi.

Let's look at some quantum algorithms, starting with the Deutsch–Jozsa algorithm.

Deutsch–Jozsa Algorithm

The Deutsch–Jozsa algorithm was designed by David Deutsch and Richard Jozsa in 1992. This algorithm was modified by Richard Cleve and others in 1998. This method is nondeterministic and created a barrier between the quantum and classical toughness of the problem. This technique is about quantum amplitudes having positive and negative values. The algorithm identifies the black-box oracle function. Figure 4-11 shows an example of an oracle function.

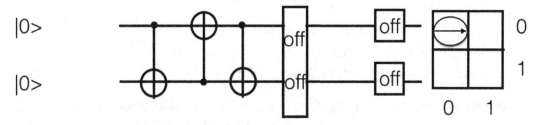

Figure 4-11. *Oracle function*

This algorithm uses a black-box (Oracle) function that takes as input the binary values and gives either 0 or 1 as output. The method finds whether the function g used is constant or balanced.

In the next sections, we discuss the algorithm implementation. To start with, let's look at the quantum register. The quantum register has a constructor and GetMeasure and ApplyOperation methods. The GetMeasure method computes the quantum states and probabilities. The ApplyOperation method updates the class instance variable data to the product of the input gate matrix and data instance.

Code Sample

<u>**Quantum_Register.py**</u>

```python
import numpy as nump
from itertools import product
from Quantum_Gate import Quantum_Gate

class Quantum_Register:
    def __init__(self, n_qbits, init):
        self._n = n_qbits
        assert len(init) == self._n

        self._data = nump.zeros((2 ** self._n), dtype=nump.complex64)
        self._data[int('0b' + init, 2)] = 1

    def Get_Measure(self):
        probs = nump.real(self._data) ** 2 + nump.imag(self._data) ** 2
        states = nump.arange(2 ** self._n)
        mstate = nump.random.choice(states, size=1, p=probs)[0]
        return f'{mstate:>0{self._n}b}'
```

```
def Apply_Gate(self, gate):
    assert isinstance(gate, Quantum_Gate)
    assert self._n == gate._n
    self._data = gate._data @ self._data
```

Now let's look at the quantum gate implementation. It has a constructor, matrix multiplication, and power methods. The matrix multiplication method operates on the input data and the instance data to return a Kronecker product. The power method takes the input to return the quantum gate.

The I, H, X, Y, and Z gates are initialized in this class. The U_Quantum_Gate gate method creates a quantum gate based on the input parameters.

Code Sample

Quantum_Gate.py

```
import numpy as nump
from itertools import product

class Quantum_Gate:
    def __init__(self, matrix):
        self._data = nump.array(matrix, dtype=nump.complex64)

        assert len(self._data.shape) == 2
        assert self._data.shape[0] == self._data.shape[1]

        self._n = nump.log2(self._data.shape[0])

        assert self._n.is_integer()

        self._n = int(self._n)

    def __matmul__(self, other):
        return Quantum_Gate(nump.kron(self._data, other._data))

    def __pow__(self, n, modulo=None):
        x = self._data.copy()

        for _ in range(n - 1):
            x = nump.kron(x, self._data)

        return Quantum_Gate(x)
```

```
IGate = Quantum_Gate([[1, 0], [0, 1]])
HGate = Quantum_Gate(nump.array([[1, 1], [1, -1]]) / nump.sqrt(2))
XGate = Quantum_Gate([[0, 1], [1, 0]])
YGate = Quantum_Gate([[0, -1j], [1j, 0]])
ZGate = Quantum_Gate([[1, 0], [0, -1]])

def U_Quantum_Gate(f, n):
    m = n + 1

    U = nump.zeros((2**m, 2**m), dtype=nump.complex64)

    def bin2int(xs):
        r = 0
        for i, x in enumerate(reversed(xs)):
            r += x * 2 ** i
        return r

    for xs in product({0, 1}, repeat=m):
        x = xs[:~0]
        y = xs[~0]

        z = y ^ f(*x)

        instate = bin2int(xs)
        outstate = bin2int(list(x) + [z])
        U[instate, outstate] = 1

    return Quantum_Gate(U)
```

Functions g1_func, g2_func, g3_func, and g4_func are shown in the code sample. g1_func takes v and returns v. g2_func takes v as an input value and returns value 1. g3_func takes v and w as input values and returns the binary XOR of v and w. g4_func takes v, w, and x as input values and returns a zero value. The CheckIfConstant method takes the input function and an integer to validate whether the function is a constant or balanced function.

Code Sample

Deutsch_Jozsa_Algorithm.py

```python
from Quantum_Register import Quantum_Register
from Quantum_Gate import HGate, IGate, UGate

def Check_If_Constant(g, n):
    qr = Quantum_Register(n + 1, '0' * n + '1')
    qr.Apply_Gate(HGate ** (n + 1))
    qr.Apply_Gate(UGate(g, n))
    qr.Apply_Gate(HGate ** n @ IGate)

    return qr.Get_Measure()[:~0] == '0' * n

def g1_func(v):
    return v

def g2_func(w):
    return 1

def g3_func(v, w):
    return v ^ w

def g4_func(v, w,x):
    return 0

print('g(v) = v is {}'.format('constant function' if Check_If_Constant
(g1_func, 1) else 'balanced function'))
print('g(v) = 1 is {}'.format('constant function' if Check_If_Constant
(g2_func, 1) else 'balanced function'))
print('g(v, w) = v ^ w is {}'.format('constant function' if Check_If_
Constant(g3_func, 2) else 'balanced function'))
print('g(v, w, x) = 0 is {}'.format('constant function' if Check_If_
Constant(g4_func, 3) else 'balanced function'))
```

Command for Execution

```
pip3 install numpy
python3 Deutsch_Jozsa_Algorithm.py
```

Output

```
g(v) = v is balanced function
g(v) = 1 is constant function
g(v, w) = v ^ w is balanced function
g(v, w, x) = 0 is constant function
```

Note   A balanced Boolean function yields output of Booleans 0s equal to 1s. A constant function has an output value as a constant.

Simon's Algorithm

Now let's look at Simon's quantum algorithm. This algorithm helps in identifying a black-box function h(x), where h(x) = h(y) and x= y $\oplus$ t. Note that t $\in$ {0, 1}n and $\oplus$ represents bitwise addition based on module 2. The goal of the algorithm is to find that by using h. This quantum algorithm performs exponentially better than the classical equivalent. This algorithm inspired Shor to come up with Shor's algorithm.

Let's look at the Python code implementation of this algorithm.

The RetrieveOnetoOneMap method takes the input mask as a parameter and returns the output, which is a bitmap function. The RetrieveValidTwoToOneMap method takes input parameters such as search mask and random seed. It returns output that is a two-to-one mapping function.

Code Sample

Simons_Algorithm.py

```python
from collections import defaultdict

import numpy as nump
from mock import patch

from collections import defaultdict
from operator import xor
from typing import Dict, Tuple, List
import numpy.random as rand
```

```python
from pyquil import Program
from pyquil.api import QuantumComputer
from pyquil.gates import H, MEASURE

def RetrieveOnetoOneMap(mask: str) -> Dict[str, str]:
        n_bits = len(mask)
        form_string = "{0:0" + str(n_bits) + "b}"
        bit_map_dct = {}
        for idx in range(2**n_bits):
        bit_string = form_string.format(idx)
        bit_map_dct[bit_string] = InvokeBitwiseXorOperation(bit_string,
        mask)
        return bit_map_dct

def RetrieveValidTwotoOneMap(mask: str, random_seed: int = None) ->
Dict[str, str]:
        if random_seed is not None:
        rand.seed(random_seed)
        bit_map = RetrieveOnetoOneMap(mask)
        n_samples = int(len(bit_map.keys()) / 2)

        range_of_2to1_map = list(rand.choice(list(sorted(bit_map.keys())),
        replace=False, size=n_samples))

        list_of_bitstring_tuples = sorted([(k, v) for k, v in bit_map.
        items()], key=lambda x: x[0])

        bit_map_dct = {}
        for cnt in range(n_samples):
        bitstring_tup = list_of_bitstring_tuples[cnt]
        val = range_of_2to1_map[cnt]
        bit_map_dct[bitstring_tup[0]] = val
        bit_map_dct[bitstring_tup[1]] = val

        return bit_map_dct
```

The Simons_Algorithm class is used to find the black-box function to satisfy a set of values. It has a constructor and the following methods: RetrieveSimonCircuit, CalculateUnitaryOracleMatrix, RetrieveBitMaskRetrieveSampleIndependentBits,

RetrieveInverseMaskEquation, AppendToDictofIndepBits, RetrieveMissingMsbVector, and CheckValidMaskIfCorrect. The RetrieveSimonCircuit method takes as input the quantum bits and returns the output, which is the quantum circuit.

The CalculateUnitaryOracleMatrix method of the Simons_Algorithm class takes the input map of bit strings. The output will be a dense matrix, which is a permutation of the values from the dictionary. The RetrieveBitMask method of the Simons_Algorithm class takes as input the quantum computer and bit strings. The output is a bit string map.

The RetrieveSampleIndependentBits method takes as input the quantum computer and bit strings. The independent bit vectors are updated with the bit strings. The RetrieveInverse mask equation method creates the bit mask based on the input function. The AppendToDictofIndepBits method takes as input an array. This method adds to the dictionary of the independent vectors. The RetrieveMissingMsbVector method searches the provenance value in the set of independent bit vectors. The method modifies the collection by adding the unit vector. The CheckValidMaskIfCorrect method validates the given mask if it is correct.

Let's now look at what an independent vector is. If two vectors are not zero and are not parallel, they are linearly independent. A linear independent vector is typically a nonzero vector. The converse is also true. After the independent vector, let's see what a bit mask is. A bit mask helps in retrieving the bits in a given byte of information.

```python
class Simons_Algorithm(object):
        def __init__(self):
        self.unitary_function_mapping = None
        self.n_qubits = None
        self.n_ancillas = None
        self._qubits = None
        self.computational_qubits = None
        self.ancillas = None
        self.simon_circuit = None
        self._dict_of_linearly_indep_bit_vectors = {}
        self.search_mask = None
        self.bit_map = None
        self.classical_register = None

    def RetrieveSimonCircuit(self) -> Program:
    simon_circuit = Program()
```

```python
oracle_name = "SIMON_ORACLE_FUNCTION"
simon_circuit.defgate(oracle_name, self.unitary_function_mapping)

simon_circuit.inst([H(i) for i in self.computational_qubits])
simon_circuit.inst(tuple([oracle_name] + sorted(self._qubits,
reverse=True)))
simon_circuit.inst([H(i) for i in self.computational_qubits])
return simon_circuit

def _Initialize_Attributes(self, bitstring_map: Dict[str, str])
-> None:
self.bit_map = bitstring_map
self.n_qubits = len(list(bitstring_map.keys())[0])
self.n_ancillas = self.n_qubits
self._qubits = list(range(self.n_qubits + self.n_ancillas))
self.computational_qubits = self._qubits[:self.n_qubits]
self.ancillas = self._qubits[self.n_qubits:]
self.unitary_function_mapping, _ = self.CalculateUnitaryOracle(bi
tstring_map)
self.simon_circuit = self.RetrieveSimonCircuit()
self._dict_of_linearly_indep_bit_vectors = {}
self.search_mask = None

@staticmethod
 def CalculateUnitaryOracle(bitstring_map: Dict[str, str]) ->
 Tuple[nump.ndarray,Dict[str, str]]:
n_bits = len(list(bitstring_map.keys())[0])

ufunc = nump.zeros(shape=(2 ** (2 * n_bits), 2 ** (2 * n_bits)))
index_mapping_dct = defaultdict(dict)
for b in range(2**n_bits):
pad_str = nump.binary_repr(b, n_bits)
for k, v in bitstring_map.items():
        index_mapping_dct[pad_str + k] =
        InvokeBitwiseXorOperation(pad_str, v) + k
        i, j = int(pad_str+k, 2), int(InvokeBitwiseXorOperation(p
        ad_str, v) + k, 2)
```

```
        ufunc[i, j] = 1
    return ufunc, index_mapping_dct

    def RetrieveBitMask(self, qc: QuantumComputer, bitstring_map:
    Dict[str, str]) -> str:
    self._Initialize_Attributes(bitstring_map)

    self.RetrieveSampleIndependentBits(qc)
    self.RetrieveInverseMaskEquation()

     if self.CheckValidMaskIfCorrect():
     return self.search_mask
     else:
     raise Exception("No valid mask found")

    def RetrieveSampleIndependentBits(self, quantum_computer:
    QuantumComputer) -> None:
    while len(self._dict_of_linearly_indep_bit_vectors) < self.n_
    qubits - 1:

     prog = Program()
     simon_ro = prog.declare('ro', 'BIT', len(self.computational_
     qubits))
     prog += self.simon_circuit
     prog += [MEASURE(qubit, ro) for qubit, ro in zip(self.
     computational_qubits, simon_ro)]
        executable = quantum_computer.compile(prog)
     sampled_bit_string = nump.array(quantum_computer.run(executable)
     [0], dtype=int)

    self.AppendToDictofIndepBits(sampled_bit_string)

     def RetrieveInverseMaskEquation(self) -> None:
     missing_msb = self.RetrieveMissingMsbVector()
     upper_triangular_matrix = nump.asarray(
     [tup[1] for tup in sorted(zip(self._dict_of_linearly_indep_bit_
     vectors.keys(),
```

```
                                   self._dict_of_linearly_indep_
                                      bit_vectors.values()),
                                key=lambda x: x[0])])

    msb_unit_vec = nump.zeros(shape=(self.n_qubits,), dtype=int)
    msb_unit_vec[missing_msb] = 1

    self.search_mask = RetrieveBinaryBackSubstitute(upper_triangular_
    matrix, msb_unit_vec).tolist()

def AppendToDictofIndepBits(self, z: nump.ndarray) -> None:
    if (z == 0).all() or (z == 1).all():
    return None
    msb_z = RetrieveMostSignificantBit(z)

     if msb_z not in self._dict_of_linearly_indep_bit_vectors.keys():
    self._dict_of_linearly_indep_bit_vectors[msb_z] = z
    else:
     conflict_z = self._dict_of_linearly_indep_bit_vectors[msb_z]
     not_z = [xor(conflict_z[idx], z[idx]) for idx in range(len(z))]
     if (nump.asarray(not_z) == 0).all():
         return None
     msb_not_z = most_significant_bit(nump.asarray(not_z))
     if msb_not_z not in self._dict_of_linearly_indep_bit_vectors.
     keys():
             self._dict_of_linearly_indep_bit_vectors[msb_not_z] =
             not_z

def RetrieveMissingMsbVector(self) -> int:
    missing_msb = None
    for idx in range(self.n_qubits):
    if idx not in self._dict_of_linearly_indep_bit_vectors.keys():
            missing_msb = idx

    if missing_msb is None:
    raise ValueError("Expected a missing provenance, but didn't find
    one.")
```

```python
        augment_vec = numpy.zeros(shape=(self.n_qubits,))
        augment_vec[missing_msb] = 1
        self._dict_of_linearly_indep_bit_vectors[missing_msb] = augment_
        vec.astype(int).tolist()
        return missing_msb

    def CheckValidMaskIfCorrect(self) -> bool:
        mask_str = ''.join([str(b) for b in self.search_mask])
        return all([self.bit_map[k] == self.bit_map[InvokeBitwiseXorOperat
        ion(k, mask_str)] for k in self.bit_map.keys()])


PADDED_BINARY_BIT_STRING = "{0:0{1:0d}b}"


def CheckValidIfUnitary(matrix: numpy.ndarray) -> bool:
        rows, cols = matrix.shape
        if rows != cols:
        return False
        return numpy.allclose(numpy.eye(rows), matrix.dot(matrix.T.conj()))


def RetrieveMostSignificantBit(lst: numpy.ndarray) -> int:
        return numpy.argwhere(numpy.asarray(lst) == 1)[0][0]


def InvokeBitwiseXorOperation(bs0: str, bs1: str) -> str:
        if len(bs0) != len(bs1):
        raise ValueError("Bit strings are not of equal length")
        n_bits = len(bs0)
        return PADDED_BINARY_BIT_STRING.format(xor(int(bs0, 2),
        int(bs1, 2)), n_bits)


def RetrieveBinaryBackSubstitute(W: numpy.ndarray, s: numpy.ndarray) -> numpy.
ndarray:
        m = numpy.copy(s)
        n = len(s)
        for row_num in range(n - 2, -1, -1):
        row = W[row_num]
```

```python
            for col_num in range(row_num + 1, n):
            if row[col_num] == 1:
                    m[row_num] = xor(s[row_num], s[col_num])

        return m[::-1]

search_mask = '110'
bm = RetrieveValidTwotoOneMap(search_mask, random_seed=42)
expected_map = {
        '000': '001',
        '001': '101',
        '010': '000',
        '011': '111',
        '100': '000',
        '101': '111',
        '110': '001',
        '111': '101'
}
for k, v in bm.items():
        assert v == expected_map[k]

reverse_bitmap = defaultdict(list)
for k, v in bm.items():
        reverse_bitmap[v].append(k)

expected_reverse_bitmap = {
        '001': ['000', '110'],
        '101': ['001', '111'],
        '000': ['010', '100'],
        '111': ['011', '101']
}

for k, v in reverse_bitmap.items():
        assert sorted(v) == sorted(expected_reverse_bitmap[k])

with patch("pyquil.api.QuantumComputer") as quantum_computer:
        quantum_computer.run.side_effect = [
        (nump.asarray([0, 1, 1], dtype=int), ),
```

```
        (nump.asarray([1, 1, 1], dtype=int), ),
        (nump.asarray([1, 1, 1], dtype=int), ),
        (nump.asarray([1, 0, 0], dtype=int), ),
    ]

simon_algo = Simons_Algorithm()
result_mask = simon_algo.RetrieveBitMask(quantum_computer, bm)
print("mask", search_mask," result mask",result_mask)
```

Command for Execution

```
pip3 install numpy
pip3 install pyquil
python3 Simons_Algorithm.py
```

Output

```
mask 110  result mask [1, 1, 0]
```

Shor's Algorithm

Peter Shor was the first to come up with a quantum algorithm for the factorization of integers in 1994. Shor's algorithm performs in polynomial time to factor an integer M. The order of performance is O(logN). This outperforms the classical equivalent. In addition, this algorithm has the potential to break the RSA cryptographic method.

The algorithm takes a number M and finds an integer q between 1 and M, which divides M. The algorithm has two steps: the reduction of factoring to order the finding method and the order finding technique.

Let's look at the classical solution. The greatest common divisor is found for b < M . gcd(b,M). This is computed using the Euclidean method. The steps are as follows:

- If gcd(b,M) != 1, there are nontrivial factors for M, and the problem is solved.

- If gcd(b,M) == 1 , find the period of g(x) = ax mod M to find r for which g(x+r) = g(x) (this is step 1).

- If r is odd, go back to step 1.

- If ar/2 = -1 (mod M), go back to step 1.

- gcd(ar/2 +- 1, M) is a nontrivial factor of M. The problem is solved.

Now let's look at the quantum algorithm. The quantum circuit for the algorithm is based on M and a in g(x) = ax mod M: Given M, Q= 2q and Q/r > M. The input and output quantum bits have superpositions of values from 0 to Q-1, and they have q quantum bits each. The g(x) function is used to transform the states of the quantum bits. The final state will be a superposition of multiple Q states, which will be Q^2 states.

Let's look at the Shor's algorithm implementation in Python.

Shor's algorithm is used for prime factorization. The QuantumMap class has the properties of state and amplitude. QuantumEntanglement has the properties of amplitude, register, and entangled. The UpdateEntangled method of the QuantumEntanglement class takes input parameters such as state and amplitude. The RetrieveEntangles method returns the list of entangled states.

Code Sample

Shors Algorithm.py

```
class QuantumMap:
    def __init__(self, state, amplitude):
        self.state = state
        self.amplitude = amplitude

class QuantumEntanglement:
    def __init__(self, amplitude, register):
        self.amplitude = amplitude
        self.register = register
        self.entangled = {}

    def UpdateEntangled(self, fromState, amplitude):
        register = fromState.register
        entanglement = QuantumMap(fromState, amplitude)
```

```
        try:
                self.entangled[register].append(entanglement)
            except KeyError:
                self.entangled[register] = [entanglement]

    def RetrieveEntangles(self, register = None):
            entangles = 0
            if register is None:
                for states in self.entangled.values():
                        entangles += len(states)
            else:
                    entangles = len(self.entangled[register])

        return entangles
```

The QuantumRecord class has numBits, numStates, an entangled list, and a states array. The SetPropagate property of the QuantumRecord class takes fromRegister as the parameter and sets propagate on the register. The UpdateMap method takes such inputs as toRegister, mapping, and propagate parameters. This method sets the normalized tensorX and tensorY lists. The FindMeasure method of the QuantumRecord class returns the output, which is the final X state. The RetrieveEntangles method takes as input the register and returns the output, which is the entangled state value. The RetrieveAmplitudes method returns the output, which is the amplitudes array of quantum states.

Code Sample

```
class QuantumRecord:
    def __init__(self, numBits):
        self.numBits = numBits
        self.numStates = 1 << numBits
        self.entangled = []
        self.states = [QuantumEntanglement(complex(0.0), self) for x in
        range(self.numStates)]
        self.states[0].amplitude = complex(1.0)

    def UpdatePropagate(self, fromRegister = None):
        if fromRegister is not None:
```

```
        for state in self.states:
            amplitude = complex(0.0)

            try:
                entangles = state.entangled[fromRegister]
                for entangle in entangles:
                    amplitude += entangle.state.amplitude *
                    entangle.amplitude

                state.amplitude = amplitude
            except KeyError:
                state.amplitude = amplitude

    for register in self.entangled:
        if register is fromRegister:
            continue

        register.UpdatePropagate(self)

def UpdateMap(self, toRegister, mapping, propagate = True):
    self.entangled.append(toRegister)
    toRegister.entangled.append(self)

    mapTensorX = {}
    mapTensorY = {}
    for x in range(self.numStates):
        mapTensorX[x] = {}
        codomain = mapping(x)
        for element in codomain:
            y = element.state
            mapTensorX[x][y] = element

            try:
                mapTensorY[y][x] = element
            except KeyError:
                mapTensorY[y] = { x: element }

    def UpdateNormalize(tensor, p = False):
        lSqrt = math.sqrt
```

```
        for vectors in tensor.values():
            sumProb = 0.0
            for element in vectors.values():
                amplitude = element.amplitude
                sumProb += (amplitude * amplitude.conjugate()).
                real

            normalized = lSqrt(sumProb)
            for element in vectors.values():
                element.amplitude = element.amplitude / normalized

    UpdateNormalize(mapTensorX)
    UpdateNormalize(mapTensorY, True)

    for x, yStates in mapTensorX.items():
        for y, element in yStates.items():
            amplitude = element.amplitude
            toState = toRegister.states[y]
            fromState = self.states[x]
            toState.UpdateEntangled(fromState, amplitude)
            fromState.UpdateEntangled(toState, amplitude.
            conjugate())

    if propagate:
        toRegister.UpdatePropagate(self)

def RetrieveMeasure(self):
    measure = random.random()
    sumProb = 0.0

    # Pick a state
    finalX = None
    finalState = None
    for x, state in enumerate(self.states):
        amplitude = state.amplitude
        sumProb += (amplitude * amplitude.conjugate()).real
```

```
            if sumProb > measure:
                finalState = state
                finalX = x
                break

    if finalState is not None:
        for state in self.states:
            state.amplitude = complex(0.0)

        finalState.amplitude = complex(1.0)
        self.UpdatePropagate()

    return finalX

def RetrieveEntangles(self, register = None):
    entangles = 0
    for state in self.states:
        entangles += state.entangles(None)

    return entangles

def RetrieveAmplitudes(self):
    amplitudes = []
    for state in self.states:
        amplitudes.append(state.amplitude)

    return amplitudes
```

The FindListEntangles method prints the list of entangles. The FindListAmplitudes method prints the values of the amplitudes of the register. The InvokeHadamard method takes input such as lambda x and the quantum bit. The codomain array is returned as the output. The InvokeQModExp method takes the input parameters aval, exponent expval, and the modval operator value. The state is calculated using the method InvokeModExp.

The InvokeQft method takes inputs such as parameters x and the quantum bit Q. The quantum mapping of the state and amplitude is returned by the method InvokeQft. The FindPeriod method takes the inputs a and N. The period r for the function is returned as output from this method.

The ExecuteShors method takes inputs such as N, attempts, neighborhood, and numPeriods as parameters. This method applies Shor's algorithm to find the prime factors of a given number N.

Code Sample

```python
def FindListEntangles(register):
    print("Entangles: " + str(register.RetrieveEntangles()))

def FindListAmplitudes(register):
    amplitudes = register.amplitudes()
    for x, amplitude in enumerate(amplitudes):
        print('State #' + str(x) + '\'s Amplitude value: ' +
        str(amplitude))

def InvokeHadamard(x, Q):
    codomain = []
    for y in range(Q):
        amplitude = complex(pow(-1.0, RetrieveBitCount(x & y) & 1))
        codomain.append(QuantumMap(y, amplitude))

    return  codomain

def InvokeQModExp(a, exp, mod):
    state = InvokeModExp(a, exp, mod)
    amplitude = complex(1.0)
    return [QuantumMap(state, amplitude)]

def InvokeQft(x, Q):
    fQ = float(Q)
    k = -2.0 * math.pi
    codomain = []

    for y in range(Q):
        theta = (k * float((x * y) % Q)) / fQ
        amplitude = complex(math.cos(theta), math.sin(theta))
        codomain.append(QuantumMap(y, amplitude))

    return codomain
```

```python
def DeterminePeriod(a, N):
    nNumBits = N.bit_length()
    inputNumBits = (2 * nNumBits) - 1
    inputNumBits += 1 if ((1 << inputNumBits) < (N * N)) else 0
    Q = 1 << inputNumBits

    print("The period is...")
    print("Q = " + str(Q) + "\ta = " + str(a))

    inputRegister = QuantumRecord(inputNumBits)
    hmdInputRegister = QuantumRecord(inputNumBits)
    qftInputRegister = QuantumRecord(inputNumBits)
    outputRegister = QuantumRecord(inputNumBits)

    print("Registers are instantiated")
    print("Executing Hadamard on the input")

    inputRegister.UpdateMap(hmdInputRegister, lambda x: InvokeHadamard
    (x, Q), False)

    print("Hadamard operation is invoked")
    print("Mapping input register to the output")

    hmdInputRegister.UpdateMap(outputRegister, lambda x:
    InvokeQModExp(a, x, N), False)

    print("Modular exponentiation is invoked")
    print("Executing quantum Fourier transform on the output")

    hmdInputRegister.UpdateMap(qftInputRegister, lambda x:
    InvokeQft(x, Q), False)
    inputRegister.UpdatePropagate()

    print("Quantum Fourier transform is invoked")
    print("Retrieving a measurement on the output")

    y = outputRegister.RetrieveMeasure()

    print("Measuring the Output register \ty = " + str(y))
```

```
        print("Retrieving  a measurement on the periodicity")

        x = qftInputRegister.RetrieveMeasure()

        print("Measuring QFT  \tx = " + str(x))

        if x is None:
            return None

        print("Retrieving the period via continued fractions")

        r = RetrieveContinuedFraction(x, Q, N)

        print("Determined Candidate period\tr = " + str(r))

        return r

def RetrieveBitCount(x):
        sumBits = 0
        while x > 0:
            sumBits += x & 1
            x >>= 1

        return sumBits

def RetrieveGcd(a, b):
        while b != 0:
            tA = a % b
            a = b
            b = tA

        return a

def RetrieveExtendedGCD(a, b):
        fractions = []
        while b != 0:
            fractions.append(a // b)
            tA = a % b
            a = b
            b = tA

        return fractions
```

```python
def RetrieveContinuedFraction(y, Q, N):
    fractions = RetrieveExtendedGCD(y, Q)
    depth = 2

    def RetrievePartial(fractions, depth):
        c = 0
        r = 1

        for i in reversed(range(depth)):
            tR = fractions[i] * r + c
            c = r
            r = tR

        return c

    r = 0
    for d in range(depth, len(fractions) + 1):
        tR = RetrievePartial(fractions, d)
        if tR == r or tR >= N:
            return r

        r = tR

    return r

def InvokeModExp(a, exp, mod):
    fx = 1
    while exp > 0:
        if (exp & 1) == 1:
            fx = fx * a % mod
        a = (a * a) % mod
        exp = exp >> 1

    return fx

def RetrieveRandom(N):
    a = math.floor((random.random() * (N - 1)) + 0.5)
    return a
```

```python
def RetrieveNeighBorCandidates(a, r, N, neighborhood):
    if r is None:
        return None

    for k in range(1, neighborhood + 2):
        tR = k * r
        if InvokeModExp(a, a, N) == InvokeModExp(a, a + tR, N):
            return tR

    for tR in range(r - neighborhood, r):
        if InvokeModExp(a, a, N) == InvokeModExp(a, a + tR, N):
            return tR

    for tR in range(r + 1, r + neighborhood + 1):
        if InvokeModExp(a, a, N) == InvokeModExp(a, a + tR, N):
            return tR

    return None

def ExecuteShorsAlgorithm(N, attempts = 1, neighborhood = 0.0,
numPeriods = 1):

    periods = []
    neighborhood = math.floor(N * neighborhood) + 1

    print("N value is" + str(N))
    print("Neighborhood value is = " + str(neighborhood))
    print("Number of periods is = " + str(numPeriods))

    for attempt in range(attempts):
        print("\nAttempt #" + str(attempt))

        a = RetrieveRandom(N)
        while a < 2:
            a = RetrieveRandom(N)

        d = RetrieveGcd(a, N)
        if d > 1:
            print("Determined factors classically, re-attempt")
            continue
```

```
        r = DeterminePeriod(a, N)

        print("validating the candidate period, nearby values, and
        multiples")

        r = RetrieveNeighBorCandidates(a, r, N, neighborhood)

        if r is None:
            print("Period was not determined, re-attempt")
            continue

        if (r % 2) > 0:
            print("Period is odd, re-attempt")
            continue

        d = InvokeModExp(a, (r // 2), N)
        if r == 0 or d == (N - 1):
            print("Period is trivial, re-attempt")
            continue

        print("Period found\tr = " + str(r))

        periods.append(r)
        if(len(periods) < numPeriods):
            continue

        print("\n Determining  least common multiple of all periods")

        r = 1
        for period in periods:
            d = RetrieveGcd(period, r)
            r = (r * period) // d

        b = InvokeModExp(a, (r // 2), N)
        f1 = RetrieveGcd(N, b + 1)
        f2 = RetrieveGcd(N, b - 1)

        return [f1, f2]

    return None
```

```
results = ExecuteShorsAlgorithm(35, 20, 0.01, 2)
print("Results are:\t" + str(results[0]) + ", " + str(results[1]))
```

Command for Execution

```
pip3 install numpy
python3 Shors_Algorithm.py
```

Output

```
N = 35
Neighborhood = 1
Number of periods = 2

Attempt #0
Printing the period...
Q = 2048     a = 4
Registers created
Performing Hadamard on input register
Hadamard operation applied
Mapping input register to output register
Modular exponentiation applied
Applying quantum Fourier transform on output register
Quantum Fourier transform applied
Getting a measurement on the output register
Measuring Output register      y = 9
Getting a measurement on the periodicity register
Measuring QFT register      x = 1024
Getting the period via continued fractions
Traceback (most recent call last):
  File "Shors_Algorithm.py", line 384, in <module>
    results = ExecuteShorsAlgorithm(35, 20, 0.01, 2)
  File "Shors_Algorithm.py", line 343, in ExecuteShorsAlgorithm
    r = DeterminePeriod(a, N)
  File "Shors_Algorithm.py", line 230, in DeterminePeriod
    r = RetrieveContinuedFraction(x, Q, N)
  File "Shors_Algorithm.py", line 281, in RetrieveContinuedFraction
```

```
    tR = RetrievePartial(fractions, d)
NameError: name 'RetrievePartial' is not defined
apples-MacBook-Air:chapter4 bhagvan.kommadi$ python3 Shors_Algorithm.py
N value is35
Neighborhood value is = 1
Number of periods is = 2

Attempt #0
The period is...
Q = 2048      a = 24
Registers are instantiated
Executing Hadamard on the input
Hadamard operation is invoked
Mapping input register to the output
Modular exponentiation is invoked
Executing quantum Fourier transform on the output
Quantum Fourier transform is invoked
Retrieving a measurement on the output
Measuring the Output register      y = 1
Retrieving  a measurement on the periodicity
Measuring QFT      x = 1707
Retrieving the period via continued fractions
Determined Candidate period      r = 6
validating the candidate period, nearby values, and multiples
Period is trivial, re-attempt

Attempt #1
Determined factors classically, re-attempt

Attempt #2
The period is...
Q = 2048      a = 23
Registers are instantiated
Executing Hadamard on the input
Hadamard operation is invoked
Mapping input register to the output
Modular exponentiation is invoked
```

Executing quantum Fourier transform on the output
Quantum Fourier transform is invoked
Retrieving a measurement on the output
Measuring the Output register   y = 16
Retrieving a measurement on the periodicity
Measuring QFT   x = 1196
Retrieving the period via continued fractions
Determined Candidate period   r = 12
validating the candidate period, nearby values, and multiples
Period found   r = 12

Attempt #3
The period is...
Q = 2048   a = 2
Registers are instantiated
Executing Hadamard on the input
Hadamard operation is invoked
Mapping input register to the output
Modular exponentiation is invoked
Executing quantum Fourier transform on the output
Quantum Fourier transform is invoked
Retrieving a measurement on the output
Measuring the Output register   y = 4
Retrieving a measurement on the periodicity
Measuring QFT   x = 1365
Retrieving the period via continued fractions
Determined Candidate period   r = 3
validating the candidate period, nearby values, and multiples
Period was not determined, re-attempt

Attempt #4
The period is...
Q = 2048   a = 19
Registers are instantiated
Executing Hadamard on the input
Hadamard operation is invoked

Mapping input register to the output
Modular exponentiation is invoked
Executing quantum Fourier transform on the output
Quantum Fourier transform is invoked
Retrieving a measurement on the output
Measuring the Output register   y = 1
Retrieving a measurement on the periodicity
Measuring QFT   x = 683
Retrieving the period via continued fractions
Determined Candidate period   r = 3
validating the candidate period, nearby values, and multiples
Period is trivial, re-attempt

Attempt #5
The period is...
Q = 2048   a = 29
Registers are instantiated
Executing Hadamard on the input
Hadamard operation is invoked
Mapping input register to the output
Modular exponentiation is invoked
Executing quantum Fourier transform on the output
Quantum Fourier transform is invoked
Retrieving a measurement on the output
Measuring the Output register   y = 29
Retrieving a measurement on the periodicity
Measuring QFT   x = 1024
Retrieving the period via continued fractions
Determined Candidate period   r = 2
validating the candidate period, nearby values, and multiples
Period found   r = 2

 Determining least common multiple of all periods
Results are: 1, 35

Grover's Algorithm

Grover's algorithm is used for searching a database. The search algorithm can find an entry in a database of M entries with $O(\sqrt{N})$ time and $O(logN)$ space. This method was found by Lov Grover in 1996. He found that his method gave a quadratic speedup compared to other quantum algorithms. In addition, Grover's method gives exponential speed over the classical equivalents.

Grover's algorithm inverts a function. Let's say $y = g(x)$. Grover's algorithm finds x when y is given as input. Searching for a value of y in the database is similar to finding a match of x. It is popular when solving NP-hard problems by using exhaustive search over a possible set of solutions. You can use Grover's algorithm to search for a phone number in a phone book of M entries in $\sqrt{M}$ steps.

Grover's algorithm consists of the following steps:

1. Initialize the system to the state.

2. Execute the Grover iteration r(M) ties.

3. Measure the amplitude Ω.

4. From Ωw (result), find w.

Let's look at the implementation of Grover's algorithm in Python.

Grover's search algorithm searches the target value of a group by computing the mean amplitude and Grover's amplitude. The plot of the graph is computed from the amplitudes derived from Grover's algorithm. The plot presents the target value with the highest amplitude. The ApplyOracleFunction method takes input x and returns the output as the hex digest of x. ApplyGrover's algorithm takes as input the target, objects, nvalue, and rounds. The amplitude is returned as the output.

The target of the algorithm is to search for 9 in the group of string objects {'14', '5', '13', '7','9','11','97'}. The amplitude is searched from the dictionary (group of string objects) based on the value 1/ square root of the length of the set (9).

Code Sample

Grovers_Algorithm.py

```
import matplotlib.pyplot as plot
import numpy as nump
import string
```

```python
import hashlib
from math import sqrt, pi
from collections import OrderedDict
from statistics import mean

def RenderGraph(amplitude_value, n):
    y_position = nump.arange(n)
    plot.bar(y_position, amplitude_value.values(), align='center', color='g')
    plot.xticks(y_position, amplitude_value.keys())
    plot.ylabel('Amplitude Value')
    plot.title('Grovers Algorithm')
    plot.show()

def ApplyOracleFunction(xvalue):
    return hashlib.sha256(bytes(xvalue, 'utf-8')).hexdigest()

def ApplyGroverAlgorithm(target, objects, nvalue, rounds):
    y_pos = nump.arange(nvalue)
    amplitude = OrderedDict.fromkeys(objects, 1/sqrt(nvalue))

    for i in range(0, rounds, 2):
        for k, v in amplitude.items():
            if ApplyOracleFunction(k) == target:
                amplitude[k] = v * -1

        average = mean(amplitude.values())
        for k, v in amplitude.items():
            if ApplyOracleFunction(k) == target:
                amplitude[k] = (2 * average) + abs(v)
                continue
            amplitude[k] = v-(2*(v-average))
    return amplitude

target_algorithm = '9'
objects_grover = ('14', '5', '13', '7','9','11','97')
number = len(objects_grover)
amplitude_grover = OrderedDict.fromkeys(objects_grover, 1/sqrt(number))
```

```
amplitude_grover[target_algorithm] = amplitude_grover[target_algorithm] * -1
print(amplitude_grover)
average_grover = mean(amplitude_grover.values())
print("Mean is {}".format(average_grover))
for k, v in amplitude_grover.items():
    if k == target_algorithm:
        amplitude_grover[k] = (2 * average_grover) + abs(v)
        continue
    amplitude_grover[k] = v-(2*(v-average_grover))
print(amplitude_grover)

needle_value = "2d711642b726b04401627ca9fbac32f5c8530fb1903cc4db02258
717921a4881"
haystack_value = string.ascii_lowercase
num = len(haystack_value)
num_rounds = int((pi / 4) * sqrt(num))
print("number of rounds are {}".format(num_rounds))
RenderGraph(ApplyGroverAlgorithm(needle_value, haystack_value, num, num_
rounds), num)
```

Command for Execution

```
pip3 install numpy
python3 Grovers_Algorithm.py
```

Output

```
OrderedDict([('4', 0.3779644730092272), ('5', 0.3779644730092272), ('3',
-0.3779644730092272), ('7', 0.3779644730092272), ('9', 0.3779644730092272),
('11', 0.3779644730092272), ('97', 0.3779644730092272)])
Mean is 0.26997462357801943
OrderedDict([('4', 0.16198477414681167), ('5', 0.16198477414681167),
('3', 0.9179137201652661), ('7', 0.16198477414681167), ('9',
0.16198477414681167), ('11', 0.16198477414681167), ('97',
0.16198477414681167)])
number of rounds are 4
```

Quantum Subroutines

Quantum computer simulators are built to simulate a quantum computer of size n quantum bits. Quantum computer simulation packages are now available to write software related to quantum computing. The packages are like black boxes that abstract the methods and techniques of quantum algorithms. The quantum Fourier transform, QPhase, and matrix inversion subroutines are available in the packages to help solve complex problems in real life.

The software packages are Project Q, IBM Quantum Experience, PennyLane, QuidPro, Liquid, etc. Quantum circuits can be simulated using the packages for analysis and computation.

The applications are developed using the quantum subroutine packages for quantum and classical algorithms. The code is compiled, and the input files are used for quantum algorithms to execute the algorithm. Quantum circuits are modeled using the subroutine API packages. The quantum circuits have a set of control gates, and measurements are performed on the simulated quantum circuits.

Execution of the quantum algorithms takes place using the group of quantum circuits that represent and model the real-life problem. The postprocessing of the results is performed for analysis purposes.

Summary

In this chapter, we looked at quantum circuits, quantum communication, quantum noise, quantum error correction, and the limitations of quantum computing. Quantum algorithms such as Deutsch–Jozsa, Simon, Shor's, and Grover's were discussed in detail. Quantum subroutines and algorithms were also presented with examples.

CHAPTER 5

Quantum Simulators

Introduction

"The most important application of quantum computing in the future is likely to be a computer simulation of quantum systems, because that's an application where we know for sure that quantum systems in general cannot be efficiently simulated on a classical computer."

—David Deutsch

This chapter gives an overview of quantum simulators. You will see how quantum simulators are used in real life. Code examples are presented for the quantum simulators, and the quantum languages and platforms are discussed in detail in this chapter.

Initial Setup

You need to set up Python 3.5 to run the code samples in this chapter. You can download it from https://www.python.org/downloads/release/python-350/.

Quantum Languages

Quantum algorithms such as Shor's factoring, Grover's search, and quantum approximate optimization are used to develop real-life applications, and quantum computers are simulated to solve real-life problems with the applications. When a quantum computing device is simulated, the software application is written in a quantum language, and the application code gets executed using the simulator.

© Bhagvan Kommadi 2020
B. Kommadi, *Quantum Computing Solutions*, https://doi.org/10.1007/978-1-4842-6516-1_5

A quantum language is a typical programming language that has quantum instructions (see Figure 5-1). The quantum instructions operate on the quantum gates in a quantum circuit. The quantum circuit is modeled in the quantum simulator.

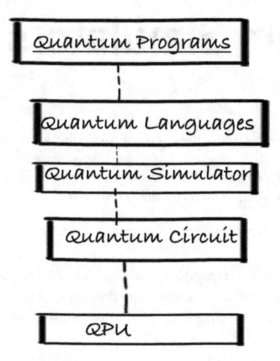

Figure 5-1. *Quantum languages*

A quantum language is a language, like C, that has data types and methods. These data types and methods are similar to a classical computer language. In addition, quantum data types and methods are provided in the language.

Quantum computers are simulated on classical machines, so a quantum simulator has features to execute tasks on a classical machine. The simulator is based on the quantum mechanics principles and quantum computer model. Yuri Manin and Richard Feynman came up with a hypothesis that simulations can be executed based on the quantum mechanical principles. A quantum computer is based on the trapped ion approach and superconducting quantum bits.

A trapped ion quantum computer is based on the confined ions suspended in electromagnetic fields. Quantum bits are stored in each ion. An ion has electronic states that have the quantum bits. The ion's quantized motion is used to shift the quantum information by using the Coulomb force. A superconducting quantum bit is the fundamental unit of a quantum computer.

Quantum computers operate using the quantum principles and use the subatomic particle state to manage the quantum information. Superposition is a quantum state that represents the subatomic particle probable state. Adiabatic quantum computation is another approach for a quantum computer that is based on the adiabatic theorem. This mathematical approach is used to execute quantum computing tasks.

IBM, Google, Microsoft, and Rigetti quantum simulators have a programming API, software development kit, and cloud services. A quantum simulator helps programmers to execute simulations and computing tasks (see Figure 5-2).

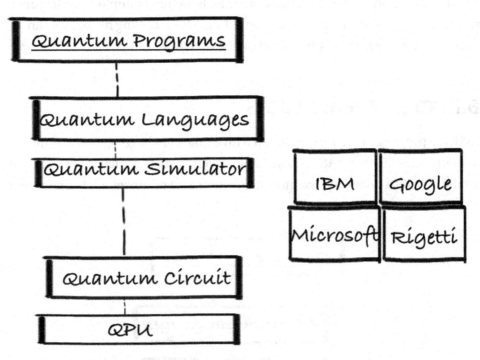

Figure 5-2. *Quantum simulators*

Quantum simulators can be executed on a classical computer. Commercial quantum computers are available, but the number of quantum bits is the difference between the different implementations.

Note   A quantum computer cannot violate the Church-Turing thesis. Hence, it is possible to have a quantum simulator that is based on a Turing machine.

Qubit Measurement

Quantum computing tasks have encoding information and quantum measuring subtasks. The encoding is based on the quantum algorithm used to solve a real-life problem. Quantum bits that are in different states such as superposition need to be measured during the quantum computing tasks.

Quantum bit measurement creates an output that is a superposition state. The Born rule states that the output will be a superposition state, and the output is proportional to the squared coefficient of the superposition state. Repetitive sampling is a feature of the software development kits in quantum computing. The sampling histograms are created based on the frequencies that are derived on the outcome counts.

Quantum Instruction Sets

The software programs that are developed with a quantum language are compiled using the quantum compiler to create a quantum instruction set (see Figure 5-3). The compiler creates the code in a quantum assembly language. Code optimization is a compiler feature.

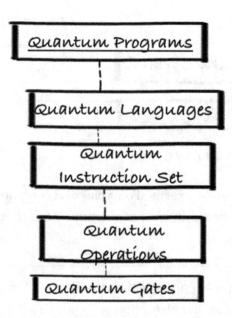

Figure 5-3. *Quantum instruction set*

A quantum assembly language helps to model quantum circuits using the quantum gates and quantum operations. A quantum processor helps to execute the instruction sets on the quantum computer. The quantum computation needs to be noise free and error free. Reliable quantum bits and fault-tolerant quantum computing methods are used for quantum computation.

Note   A quantum instruction set is similar to the classical instruction sets that are present in classical computers.

Code Sample

```
import cirq

qbit = cirq.GridQubit(0, 0)

quantum_circuit = cirq.Circuit.from_ops(
    cirq.X(qbit)**0.5,
    cirq.measure(qbit, key='m')
)
print("Quantum Circuit:", quantum_circuit)
quantum_simulator = cirq.Simulator()
results = quantum_simulator.run(quantum_circuit, repetitions=20)
print("Results ",results)
```

Command for Execution

```
pip3 install cirq
python3 quantum_circuit.py
```

Output

```
Quantum Circuit: (0, 0): ——X^0.5——M('m')——
Results  m=11000011010011101100
```

Let's now look at a full-stack universal quantum computer. A full-stack universal quantum computer is based on the classical computer stack.

Full-Stack Universal Quantum Simulator

A quantum simulator is based on the Feynman conjecture of a quantum system simulated for computational tasks. Seth Lloyd presented the quantum principles that can be used for the full-stack universal quantum simulator. A full-stack universal quantum simulator is based on the concept of a quantum computational model. A quantum device is used to build a quantum computer, which can work in parallel to a classical computer. The quantum processors are used to execute any computational tasks that need a speedup. This hybrid approach uses classical and quantum computers to execute tasks. See Figure 5-4.

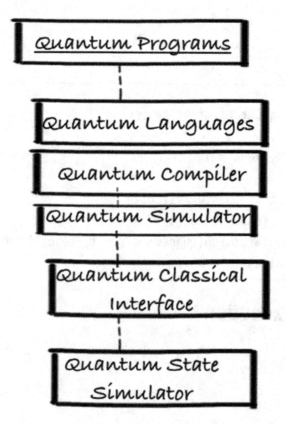

Figure 5-4. *Full-stack universal quantum simulator*

Quantum accelerators and quantum-assisted optimization are the latest techniques used for quantum computation. A full-stack quantum simulator is based on the quantum computational layers and components.

Code Sample

```python
import cirq
import random

def rand2d(rows, cols):
    return [[random.choice([+1, -1]) for _ in range(cols)] for _ in
    range(rows)]

def random_instance(length):
    h = rand2d(length, length)
    jr = rand2d(length - 1, length)
    jc = rand2d(length, length - 1)
    return (h, jr, jc)
def one_step(h, jr, jc, x_half_turns, h_half_turns, j_half_turns):
    length = len(h)
    yield rot_x_layer(length, x_half_turns)
    yield rot_z_layer(h, h_half_turns)
    yield rot_11_layer(jr, jc, j_half_turns)
def rot_x_layer(length, half_turns):
    rot = cirq.XPowGate(exponent=half_turns)
    for i in range(length):
        for j in range(length):
            yield rot(cirq.GridQubit(i, j))
def rot_z_layer(h, half_turns):
    gate = cirq.ZPowGate(exponent=half_turns)
    for i, h_row in enumerate(h):
        for j, h_ij in enumerate(h_row):
            if h_ij == 1:
                yield gate(cirq.GridQubit(i, j))
def rot_11_layer(jr, jc, half_turns):
    gate = cirq.CZPowGate(exponent=half_turns)
    for i, jr_row in enumerate(jr):
        for j, jr_ij in enumerate(jr_row):
            if jr_ij == -1:
                yield cirq.X(cirq.GridQubit(i, j))
                yield cirq.X(cirq.GridQubit(i + 1, j))
```

```
            yield gate(cirq.GridQubit(i, j),
                        cirq.GridQubit(i + 1, j))
            if jr_ij == -1:
                yield cirq.X(cirq.GridQubit(i, j))
                yield cirq.X(cirq.GridQubit(i + 1, j))

    for i, jc_row in enumerate(jc):
        for j, jc_ij in enumerate(jc_row):
            if jc_ij == -1:
                yield cirq.X(cirq.GridQubit(i, j))
                yield cirq.X(cirq.GridQubit(i, j + 1))
            yield gate(cirq.GridQubit(i, j),
                        cirq.GridQubit(i, j + 1))
            if jc_ij == -1:
                yield cirq.X(cirq.GridQubit(i, j))
                yield cirq.X(cirq.GridQubit(i, j + 1))

h, jr, jc = random_instance(3)
length = 3
qubits = [cirq.GridQubit(i, j) for i in range(length) for j in
range(length)]
simulator = cirq.Simulator()
circuit = cirq.Circuit()
circuit.append(one_step(h, jr, jc, 0.1, 0.2, 0.3))
circuit.append(cirq.measure(*qubits, key='x'))
results = simulator.run(circuit, repetitions=100)
print(results.histogram(key='x'))
```

Command for Execution

```
pip3 install cirq
python3 simulator.py
```

Output

```
Counter({0: 83, 8: 3, 32: 3, 4: 2, 256: 2, 1: 2, 128: 2, 16: 1, 64: 1, 257: 1})
```

Quantum Assembly Programming

Quantum computers have quantum processors that are physical chips that can have a set of quantum bits. A quantum computer has electronic units for control and components for managing memory and data. The quantum processors can be based on gate models or annealing approaches. The gate model quantum processing units have quantum circuits and computational tasks that are executed using quantum gates. Quantum annealing is used for annealing quantum processing units.

Quantum interpreters and compilers are used to compile the software code to create assembly code. A quantum assembly language is used to create the assembly code that has the quantum instruction sets that can be executed on the quantum simulator or quantum computer. See Figure 5-5.

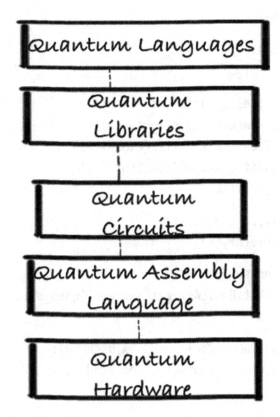

Figure 5-5. *Quantum assembly language*

Quantum physical principles are used for creating quantum computers and simulators. Programmers are provided with software development kits to help them to develop applications using the quantum programming languages.

Quantum computer architecture is based on different methods such as adiabatic computational models. Quantum circuits are developed using the quantum gates on the quantum assembly. These circuits can be simulated on the quantum simulator for executing the quantum assembly code. The code has the instruction sets to operate on the quantum gates in the circuit.

Code Sample

```python
import qiskit

class QASM_JobManager:

    def __init__(self):
        self.data = []

    def createProgram(self,quantum,classical):
        qr = qiskit.QuantumRegister(quantum)
        cr = qiskit.ClassicalRegister(classical)
        program = qiskit.QuantumCircuit(qr, cr)
        return qr,cr,program

    def measure(self,program,qr,cr):
        program.measure(qr,cr)

    def getQASMBackend(self):
        backend = qiskit.BasicAer.get_backend('qasm_simulator')
        return backend

    def executeProgramOnQASM(self,program,backend):
        job = qiskit.execute( program, backend  )
        return job
```

```
jobManager = QASM_JobManager()

quantumRegister,classicalRegister,program = jobManager.createProgram(1,1)

jobManager.measure(program,quantumRegister,classicalRegister)

backend = jobManager.getQASMBackend()

print("The device name:",backend.name())

job = jobManager.executeProgramOnQASM(program,backend)

print( job.result().get_counts() )
```

Command for Execution

```
pip3 install qiskit
python3 quantum_assembly.py
```

Output

```
The device name: qasm_simulator
{'0': 1024}
```

Quantum Hardware Platforms

Quantum hardware platforms are available from Google, Rigetti, Microsoft, and IBM (see Figure 5-6). They implement quantum computational models to perform quantum computations. The approaches used for building the quantum computer are trapped ions, photonics systems, and superconducting quantum bits. Quantum bits are the current limitation and the differentiator between these platforms.

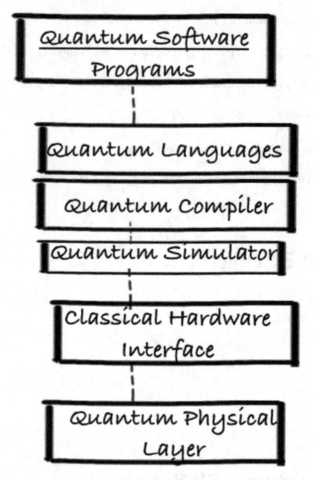

Figure 5-6. *Quantum hardware*

Quantum random access memory is based on the master classical device operating on the quantum device. This is based on the hybrid quantum programming model where you can start with typical programming languages like Python and use quantum programming languages for quantum computational tasks. In Microsoft implementations, C# is used for typical programming tasks, and Q# is used for quantum computations.

A quantum register is based on multiple quantum bit systems. It is similar to a classical computer register. A quantum register operates on quantum bits in a quantum device and is a system consisting of multiple qubits. This is equivalent to the classical processor register. Quantum computers manipulate qubits within a quantum register. It has the capability to model 2^x combinations of the quantum bits. A classical register can handle x combinations of the classical bits.

Commercially Available Quantum Computers

The commercial quantum computing platforms are listed here:

- Rigetti

- Google

- IBM

- Microsoft

The Rigetti computing platform has the following components:

- Quil

- PyQuil

- QUILC

- Quantum virtual machine (QVM)

The Google Quantum Computing Playground has support for scripting and the following features:

- Browser-based IDE

- Quantum registers of 22 qubits

- Programs and compilation

- Nested logic in programs

- Quantum gates modeling

- Math functions support

- Visualization modes

The IBM Quantum Experience has the following modules:

- Quantum Circuit Composer

- Quantum Circuit Visualizer

- Quantum Gates Support

- Quantum Register

- Qiskit Notebooks (Jupyter based)

The Microsoft Quantum Development Kit has support for Microsoft Q# and the QDK. It has a simulator and open source AI for developers to create programs using quantum circuits, gates, and simulators.

Summary

In this chapter, we looked at the quantum languages, qubit measurement, quantum instruction sets, full-stack universal quantum simulator, quantum assembly programming, and quantum hardware platforms. The chapter also presented code samples for quantum assembly, quantum circuits, and simulator.

CHAPTER 6

Quantum Optimization Algorithms

Introduction

"The mathematical framework of quantum theory has passed countless successful tests and is now universally accepted as a consistent and accurate description of all atomic phenomena."

—Erwin Schrodinger

This chapter covers an overview of quantum optimization algorithms. You will see how quantum optimization is used in real life. Code examples are presented for the quantum optimization applications.

Initial Setup

You need to set up Python 3.5 to run the code samples in this chapter. You can download it from `https://www.python.org/downloads/release/python-350/`.

Approximate Optimization Algorithms

In this section, we look at approximate algorithms such as semidefinite programming and approximate combinatorial optimization.

Approximate combinatorial optimization helps in finding a solution closer to the optimum value quickly. This technique may not give the best solution, however. This method helps in finding a short path for a traveling salesman problem. The traveling salesman problem is about a salesman who needs to travel within M cities. The salesman can visit each city once but needs to end where he started. The classic solution is to

125

© Bhagvan Kommadi 2020
B. Kommadi, *Quantum Computing Solutions*, https://doi.org/10.1007/978-1-4842-6516-1_6

set the first city as the starting and end points. The next step is to identify all possible permutations, (M-1)!, of the cities. Each permutation needs to be checked for the cost or the goal. The objective of the problem is to identify the permutation with the least cost. See Figure 6-1.

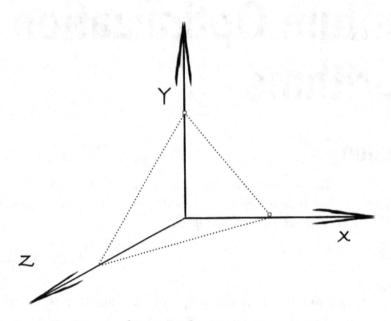

Figure 6-1. *Approximate combinatorial optimization*

Dynamic programming can be used for this problem by breaking it down into subproblems. Start at the first city, and for all other cities, create a subproblem starting from the new city. These subproblems will create a recursion to identify the cheapest travel itinerary for the salesperson.

The approximate optimization algorithm calculates the optimal itinerary by finding the cost of the minimum spanning tree. A minimum spanning tree is a subgroup of the connected path of the cities without having cycles and lesser total edge cost.

Let's now look at semidefinite programming algorithms.

Semidefinite Programming

Semidefinite programming is an important technique in optimization. The generality is the key principle behind this method. The variable matrix in the optimization problem is constrained to be positive semidefinite. The semidefinite programming term comes from the constraint related to the positive semidefiniteness.

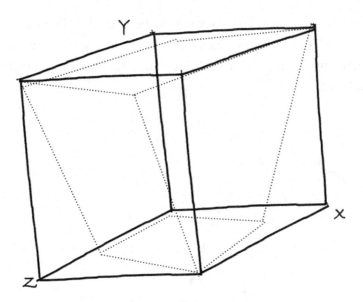

Figure 6-2. *Semidefinite programming*

The quantum semidefinite programming technique has the following steps:

1. Apply the oracle function for a search violating the constraint.

2. Generate a density matrix.

3. Calculate the phase estimation of the unitary operator on the density matrix.

    ```
    density matrix = Σk k λ k | Ψk > < Ψk| X | λ k>< λ k'|
    ```

4. Estimate $Tr[e^{\wedge}(- \text{density matrix})]$.

5. Check if the output is λ. If so, return the value $\lambda -1e^{\wedge-\lambda}$.

6. Execute step 1, and measure the density matrix of the second system.

7. If the output is λ, accept it; otherwise, reject it.

8. Keep iterating until this step is accepted.

Note   The density matrix represents the statistical state of the quantum system. In the classical system, probability distribution is the equivalent.

Semidefinite programming can be presented mathematically. Given n real numbers and c1, cn $\in$ R, and p $\times$ p Hermitian matrices D : B1,..,...,...Bp, the objective is to maximize Transpose (DX) such that Transpose(B i X) < c i X >= 0. The variable X is constrained to be positive semidefinite.

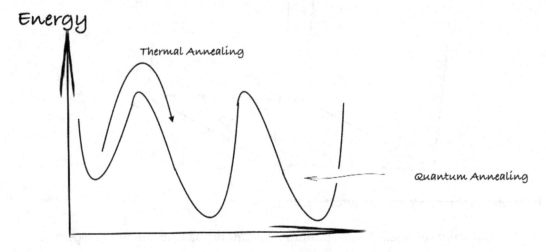

Figure 6-3. *Quantum annealing: semidefinite programming*

The problems solved by a semidefinite programming method can be with no constraints or with constraints such as equalities and inequalities. The input for the quantum semidefinite programming algorithm consists of a sparse matrix, quantum states, and quantum operators. A typical sparse matrix has at most n nonzero elements per row.

Quantum annealing is another semidefinite programming technique. It is related to the quantum system's behavior. This helps in solving NP-hard optimization problems. An optimization problem like the traveling salesman problem can be solved using quantum annealing. The solution is the ground state of the physical system, which is based on the Hamiltonian function. The Hamiltonian function describes the energy and motion of the physical system. The classical equivalent is the simulated annealing, random search, and gradient descent algorithms.

Note   The Hamiltonian operator is related to the potential and kinetic energy sum for the quantum particles. Hamiltonian is Hermitian. A Hermitian operator's eigenvalues are real, and the eigenfunctions are orthogonal. The eigen value shows the spread of the data or characteristic vector of a transformation, which is linear.

Semidefinite programming is related to using linear programming by replacing the vectors with positive semidefinite matrices. It is based on linear programming and convex quadratic programming. Semidefinite programming generalizes the linear programming by using the block diagonal, symmetric, and semidefinite matrices. These methods are used in structural design and control theory–based applications. The methods are also used in the chemical industry for identifying how to optimally mix chemical materials.

The maximum cut problem is a well-known problem where semidefinite programming is applied. The maximum cut problem is related to identifying a maximum cut in a graph. In a given network or a graph, the goal is to identify a subgroup of vertices based on the size of the edges between the vertices. The maximum cut problem is used in large-scale integration applications.

The input for the problem is the number of vertices, edges, and adjacency list with weights of the edges. Every edge has the origin and the destination. The goal is to identify the cut that has the highest total weight of the edges.

Now let's look at a semidefinite programming example in Python.

The mathematical problem is as follows:

minimize Transpose (CX)

such that Transpose(A i X) < b i X >= 0. Variable X is constrained to be positive semidefinite.

In the following Python code, CMatrix is generated randomly. The A and b matrices are also generated randomly. CVXPy has a mathematical API to solve convex optimization problems. The optimum value is calculated for the generated problem.

Code Sample

```python
import cvxpy as cvxp
import numpy as nump

n = 3
p = 3
nump.random.seed(1)
CMat = nump.random.randn(n, n)
AMat = []
bMat = []
for i in range(p):
```

```
    AMat.append(nump.random.randn(n, n))
    bMat.append(nump.random.randn())

XMat = cvxp.Variable((n,n), symmetric=True)

constraints = [XMat >> 0]
constraints += [
    cvxp.trace(AMat[i]@XMat) == bMat[i] for i in range(p)
]
problem = cvxp.Problem(cvxp.Minimize(cvxp.trace(CMat@XMat)), constraints)
problem.solve()
print("The optimal value of the problem is", problem.value)
print("A solution X is")
print(XMat.value)
```

Command

```
pip3 install cvxpy
pip3 install nump
python3 sdp_cvx.py
```

Output

The optimal value of the problem is 2.654351510231541

A solution X is

```
[[ 1.60805795 -0.59770492 -0.69576152]
 [-0.59770492  0.22229041  0.24689363]
 [-0.69576152  0.24689363  1.39679885]]
```

Similarly, real-life problems can be either maximize or minimize problems. They can be modeled as in the previous code with constraints modeled as linear equations.

QAOA

Quantum approximate optimization (QAOA) algorithms are related to identifying the solution to an optimization problem using polynomial time methods (see Figure 6-4). Typical problems solved by QAOA are NP-hard problems.

QAOA methods are executed on a quantum virtual machine. Problems such as max cut and graph partitioning are solved using QAOA. The QAOA methods have the following steps:

1. Encode the cost function using Pauli operators.

2. Model the problem using a quantum computing framework and the algorithm.

3. Calculate the ground state solution using sampling.

Tip   The Pauli operator is related to angular momentum operations. An angular momentum operation represents the spin of a quantum particle in three spatial directions.

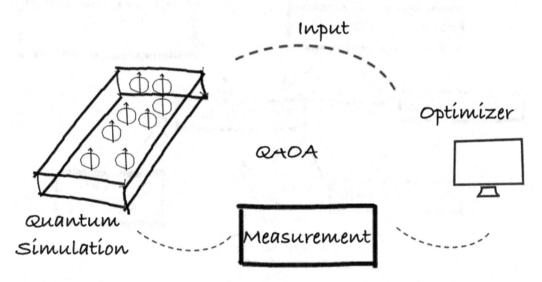

Figure 6-4. *Quantum approximate optimization*

The maximum cut problem can be solved using the QAOA method and is related to identifying a maximum cut in a graph. The ground state is modeled using a Hamiltonian function, and the state is measured. The multibody quantum state is measured, which yields a max cut solution with higher probability. In a given network or a graph, the goal is to identify a subgroup of vertices based on the size of the edges between the vertices.

Let's look at a real-life problem where QAOA can be applied (see Figure 6-5). Traffic modeling in transportation depends on traffic zones, trip tables, highway networks, vehicle capacity, and land-use models. Land-use models consist of trends, land use and zoning plans, activity areas, and maps. A transportation system has streets, highways, parking, speeds, travel times, and speed limits. A transportation network will have traffic volumes and vehicle capacities. Travel patterns identify the mode use, typical source to destination paths, and trip characteristics. Wardrop's first principle is based on network and road user balance. The journey times of all the paths used are equal and less than the unused path by one vehicle. This principle helps in minimizing the average time for a trip. There are other principles and algorithms to balance the road user and network traffic flow.

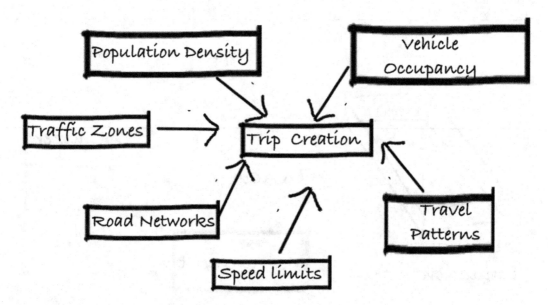

Figure 6-5. *Transportation planning: factors*

Transportation Planning: Factors

Trips are generated based on the source, destination, vehicle, and city transportation network. The network has vehicles, cyclists, and pedestrians in different time zones and space constantly moving. Transportation planning is a continuous activity, and the street network infrastructure keeps getting updated. Traffic modeling is done using techniques such as simulation and machine learning. Traffic control is done using the historical data, estimating the number of vehicles during a time period, and managing the flow of the traffic across the network.

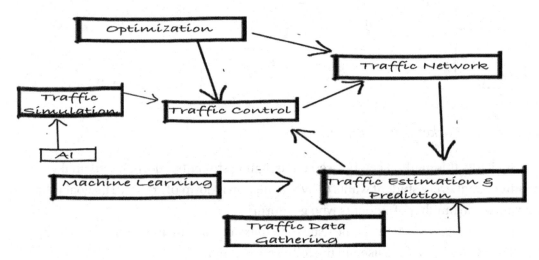

Figure 6-6. *Traffic planning: AI and machine learning*

The max flow–min cut approach is used to maximize the flow of the traffic in a transportation network. This approach is based on the maximum flow of traffic from source to destination and is the same as the minimum cut to separate the source from the destination. This approach can be applied to a computer network packet flow.

Packet flow through a network can be modeled using the max flow–min cut approach. The network has the information flow through the nodes to reach the destination. The goal is to maximize the number of packets that can flow through a computer network.

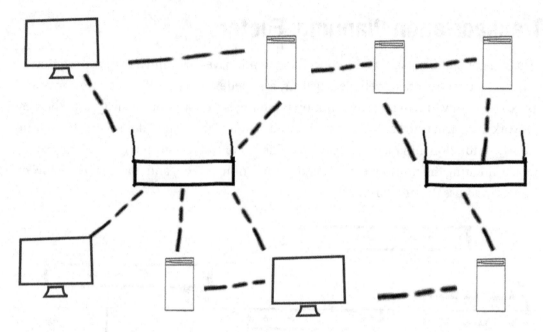

Figure 6-7. *Packet flow: computer network*

Similarly, this method can be used for assigning students a dorm based on the condition that every student wants to live in their own dorm of choice. The goal is to maximize the number of students who get dorms.

Let's look at how QAOA can be used for a max cut solution in Python. The Grove framework and PyQuil are used in the following example.

Code Sample

```
import numpy as nump
from grove.pyqaoa.maxcut_qaoa import maxcut_qaoa
import pyquil.api as pyquilapi
qvm_session = pyquilapi.QVMConnection()

configuration_ring = [(0,1),(1,2),(2,3),(3,0)]

steps = 2
inst = maxcut_qaoa(graph=configuration_ring, steps=steps)
betas, gammas = inst.get_angles()
t = nump.hstack((betas, gammas))
param_program = inst.get_parameterized_program()
```

```
program = param_program(t)
wave_function = qvm_session.wavefunction(program)
wave_function = wave_function.amplitudes

for state_index in range(inst.nstates):
    print(inst.states[state_index], nump.conj(wave_function[state_
    index])*wave_function[state_index])
```

Command

```
pip3 install nump
pip3 install grove
pip3 install pyquil
qvm  -S  (separate terminal window1)
python3 qaoa.py  (terminal window2)
```

Output

```
0000 (5.030591151913648e-10+0j)
0001 (7.986230289234977e-06+0j)
0010 (7.986230289234498e-06+0j)
0011 (2.2019121467156433e-07+0j)
0100 (7.98623028923483e-06+0j)
0101 (0.49996761419335345+0j)
0110 (2.2019121467149302e-07+0j)
0111 (7.98623028923475e-06+0j)
1000 (7.98623028923475e-06+0j)
1001 (2.2019121467149302e-07+0j)
1010 (0.49996761419335345+0j)
1011 (7.98623028923483e-06+0j)
1100 (2.2019121467156433e-07+0j)
1101 (7.986230289234498e-06+0j)
1110 (7.986230289234977e-06+0j)
1111 (5.030591151913648e-10+0j)
```

0101 and 1010 have 0.499 wave function amplitudes. These are the max cuts in the configuration. Figure 6-8 shows the 0101 configuration.

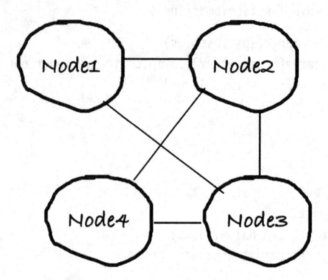

Figure 6-8. *0101 configuration*

Figure 6-9 shows the 1010 configuration.

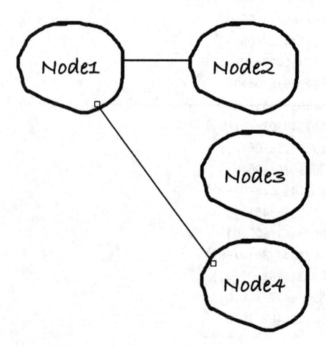

Figure 6-9. *1010 configuration*

Similarly, the traffic network and computer network can be represented as shown in the previous code. The constraints can be added using the traffic limits and network node capacities. The objective in that case will be to maximize the flow and minimize the cut.

Let's now look at the combinatorial optimization algorithms.

Combinatorial Optimization

Combinatorial optimization is used to solve real-life problems such as the traveling salesman problem and diet calculations for a person. The optimization algorithm uses possible inputs and finds the maximum or minimum output. Let's say c(x) is the goal function. If the goal is to maximize c(x) and the constraints modeled are d(x) and b(x), then we have the following:

```
d(x) < 0
b(x) =0
```

Combinatorial optimization algorithms find the solution that maximizes the goal function with discrete values. The optimization algorithm is called *linear programming* if the goal function and the constraints are linear.

Classic optimization algorithms have equivalents in quantum optimization. We'll look at the quantum semidefinite programming methods in the next section.

Now let's look at quantum nondeterministic polynomial algorithms which are based on quantum semidefinite programming.

Quantum NP (BQNP)

Bounded-error quantum polynomial time-based algorithms are the polynomial time order techniques that can be run on quantum computers. The computing knot invariants and quantum simulation algorithms are BQNP algorithms.

A good example for the BQNP algorithm is Shor's algorithm. It is based on factoring the product of prime numbers. The algorithm gets executed in polynomial time with a bounded error. The execution time depends on the number of operations in Shor's algorithm.

Shor's algorithm, named after mathematician Peter Shor, is a quantum algorithm for integer factorization. The quantum algorithm was formulated in 1994. It is related to the problem "Given an integer K, find its prime factors."

On a quantum computer, to factor an integer K, Shor's algorithm runs in polynomial time. The time taken for execution is polynomial in log K. The size of the input for the algorithm is log K. To be exact, it takes time in the order of $O((\log N)3)$. The integer factorization problem can be modeled on the quantum computer using the BQP algorithm. This algorithm is exponentially faster than the classical equivalent. The efficiency of Shor's algorithm is equal to the efficiency of the quantum Fourier transform algorithm.

Summary

In this chapter, we looked at the techniques related to the approximate optimization, semidefinite programming, combinatorial optimization, and quantum nondeterministic polynomial algorithms. In addition, we discussed approximate optimization algorithms in detail.

PART III

Quantum Solutions

## CHAPTER 7

Quantum Algorithms

Introduction

"If you can't program it, you haven't understood it."

—David Deutsch

This chapter gives you an overview of quantum algorithms. You will see how quantum algorithms are implemented in real life. Code examples are presented for the quantum algorithms, such as quantum least squares fitting and quantum sort.

Initial Setup

You need to set up Python 3.5 to run the code samples in this chapter. You can download it from `https://www.python.org/downloads/release/python-350/`.

Quantum Least Squares Fitting

In this section, we will look at the algorithm called *quantum least squares fitting*.

Quantum information theory helps in solving regression and prediction problems efficiently. Quantum algorithms that are based on the quantum mechanical principles enhance the efficiency of the least squares fitting algorithm. In quantum state tomography, the quantum states are fit to a group of functions. Quantum state tomography is related to the reconstruction of quantum states. Quantum states are measured, and an ensemble of the same quantum states is created during the process of tomography.

Tip Quantum state tomography is the method of checking the quantum state.

© Bhagvan Kommadi 2020
B. Kommadi, *Quantum Computing Solutions*, https://doi.org/10.1007/978-1-4842-6516-1_7

A continuous function is used to find the classical least squares fitting. A continuous function is continuous in every interval. In every interval, the values are continuous. Continuity has three rules, as shown here:

- The function needs to be defined at a point.

- It needs to have a limit on a side at a point.

- The limit from left or right side needs to be equal to the function value at the point.

As an example, $f(x) = x^2$ is a continuous function. The graph of the continuous function will be unbroken without any jumps and asymptotes. An *asymptote* is the point where the function jumps to infinity. A domain has a set of points that have values for a function. Let's say that $f(x)$ is a linear function of y. Then $x = A^{-1}y$, where A is a square matrix. The pseudo-inverse of A is defined as shown here:

$$A^+ = (A^TA)^{-1} A^T$$

Now, let's look at the quantum equivalent of the least squares fitting method. The quantum least squares fitting method is based on finding a simple function that is continuous and fits the discrete set of M (domain) points $\{x_j, y_j\}$. The function can be nonlinear but needs to be forced for a linear fit. The function is presented mathematically as shown here:

$$f(x, \lambda) = \Sigma_{j=1}^{M} f_j(x)\lambda_j$$

where λ_j is a component and belongs to λ and $f(x, \lambda)$.

The least squares fitting algorithm does through three steps.

1. Find the pseudo-inverse using the quantum algorithm.

2. Estimate the fitness quality using the algorithm.

3. Identify the fitness parameters λ using learning algorithm.

The fitness function measures the quality of the solution. This is done by measuring the closeness of the solution to the optimum solution. The fitness score is calculated by using the fitness parameters. The Euclidean distance and Manhattan distance methods are used to find fitness. The Euclidean distance is calculated using the sum of the squares of the perpendicular distances. The Euclidean distance for the two points (x_1, y_1) and (x_2, y_2) is as follows:

$$\sqrt{(x_1-x_2)^2 + (y_1-y_2)^2}$$

Tip   Manhattan distance is defined as the smallest distance between the point and the line. The Manhattan distance between (x_1, y_1) and $cx + dy + e = 0$ is calculated as follows: $|cx_1 + dy_1 + e| / max(|c|, |d|)$.

The algorithm takes the following inputs mentioned here:

- Vector of expected data

- Matrix, which is the basis

- Vector of weights

- Fitted matrix

- Trace of the fitted matrix

Now let's look at some of the terms that were used in the algorithm in detail.

The basis matrix has elements that are linearly independent. It is a combination of linearly independent vectors. Let's say we have a matrix B defined as shown here:

$$B = \begin{bmatrix} 1 & 0 & 0 \\ 0 & 1 & 1 \end{bmatrix}$$

Figure 7-1. *Matrix B*

The columns of the matrix B are linearly independent. The redundant column vectors are removed to find the basis of the matrix. If you try to remove the first column, the span will move to the y-axis. The basis of matrix W is as defined here using the important and pivotal columns:

$$W = \begin{bmatrix} 1 & 0 \\ 0 & 1 \end{bmatrix}$$

Figure 7-2. *Basis of Matrix B*

Vector data types are fixed-size lists. Mathematical vectors are lines of specific length with a starting point and direction. Vector examples are work, force, acceleration, and velocity. In two dimensions, the example of a vector is $b_1i + b_2j$. In three dimensions, the vector is represented as $b_1i+b_2j+b_3k$. The n dimensional vector is defined as $b_0+b_1t_1+b_2t_2+...+b_nt_n$.

Let's look at the trace of the matrix definition. The trace of a matrix is the sum of the diagonal elements from the upper left to the lower right. It is equal to the sum of the matrix's eigenvalues.

Let's say we have a matrix B defined as follows:

$$B = \begin{bmatrix} x1 & x2 & x3 \\ y1 & y2 & y3 \\ z1 & z2 & z3 \end{bmatrix}$$

Figure 7-3. *Trace of Matrix*

The trace of the matrix is defined as $x_1+y_2+z_3$.

Coming back to the algorithm, the algorithm provides a better fit for the data using quantum least squares fitting. Figure 7-4 demonstrates quantum least square fitting. The classical least square fitting algorithm might be closer to an average curve that passes through the data. Quantum least squares fitting tries to minimize the bias across different points.

Let's look into the calculation of the bias now. *Bias* is the difference between the real value and the mean value of the predicted data. The bias is because it is assumed that the linear function is a fit. Quantum least squares fitting helps to bring down the bias, as shown in Figure 7-4.

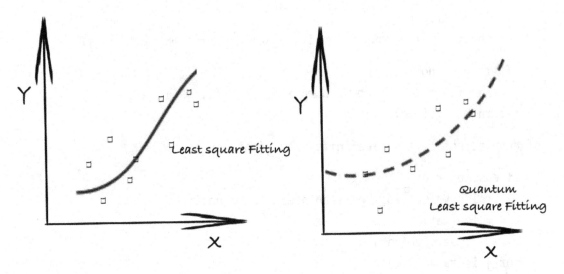

Figure 7-4. *Quantum least square fitting*

Code Sample

```
import numpy as nump
from scipy import linalg as lina
from scipy.linalg import lstsq

def get_least_squares_fit(data, basis_matrix, weights=None, PSD=False,
trace=None):

    c_mat = basis_matrix
    d_mat = nump.array(data)

    if weights is not None:
        w = nump.array(weights)
        c_mat = w[:, None] * c_mat
        d_mat = w * d_mat
    rho_fit_mat, _, _, _ = lstsq(c_mat.T, d_mat) <——- Fitting calculation
    print(rho_fit_mat)
    size = len(rho_fit_mat)
    dim = int(nump.sqrt(size))
    if dim * dim != size:
        raise ValueError("fitted vector needs to be a square matrix")
    rho_fit_mat = rho_fit_mat.reshape(dim, dim, order='F')
```

```
    if PSD is True:
        rho_fit_mat = convert_positive_semidefinite_matrix#(rho_fit)

    if trace is not None:
        rho_fit_mat *= trace / nump.trace(rho_fit_mat)
    return rho_fit_mat

def convert_positive_semidefinite_matrix(mat, epsilon=0):

    if epsilon < 0:
        raise ValueError('epsilon needs to be positive ')
    dim = len(mat)
    v, w = lina.eigh(mat)
    for j in range(dim):
        if v[j] < epsilon:
            tmp = v[j]
            v[j] = 0.
            x = 0.
            for k in range(j + 1, dim):
                x += tmp / (dim - (j + 1))
                v[k] = v[k] + tmp / (dim - (j + 1))

    matrix_psd = nump.zeros([dim, dim], dtype=complex)
    for j in range(dim):
        matrix_psd += v[j] * nump.outer(w[:, j], nump.conj(w[:, j]))

    return matrix_psd

data = [12, 21, 23.5, 24.5, 25, 33, 23, 15.5, 28,19]
u_matrix = nump.arange(0, 10)

basis_matrix = nump.array([u_matrix, nump.ones(10)])

rho_fit_val = get_least_squares_fit(data,basis_matrix)
```

Command

```
pip3 install numpy

pip3 install scipy

python3 quantum_least_squares_fit.py
```

Output

```
[ 0.45757576 20.39090909]
```

The solution is v=0.4575u+20.39.

Tip You can change the previous code for the classical and quantum methods to use it for seasonal sales data prediction: $v = w_1u+c$.

Quantum Sort

A sorting algorithm orders the elements of a group in a specified order. Classical sorting methods come in different types.

- Bubble

- Selection

- Merge

- Insertion

- Quick

- Heap

Bubble, selection, and heap methods operate by shifting the items one by one to their final position. Insertion, quick sort, and other methods place the elements in a temporary position that is near the final position, as shown in Figure 7-5.

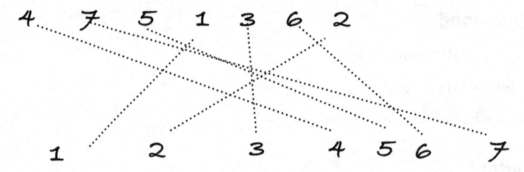

Figure 7-5. *Classical sorting method*

In real life, telephone directories and dictionaries are examples where sorting is done by name and alphabetic order, respectively. Sorting can be classified into two different techniques.

- In-place sorting

- Not-in-place sorting

In-place sorting is about shifting the items within the group and not using extra memory or space. Not-in-place sorting is about using extra space for ordering the elements. Bubble sort is based on in-place sorting, while merge sort is based on not-in-place sorting.

Let's look at the quick sort algorithm in Python. The quick sort algorithm picks the group of elements to sort them in ascending order. The input for the program is 4, 9, 18, 11, 3, 2, 0. The sorted elements are 0, 2, 3, 4, 9, 11, 18.

Code Sample

```
def sort_quick_sort_method(group):
    if len(group) <= 1:
        return group
    pivot_value = group[len(group)//2]
    left_value = [x for x in group if x < pivot_value]
    middle_value = [x for x in group if x == pivot_value]
    right_value = [x for x in group if x > pivot_value]
    return sort_quick_sort_method(left_value) + middle_value + sort_quick_
    sort_method(right_value)
print(sort_quick_sort_method([4,9,18,11,3,2,0]))
```

Command

```
python3 quick_sort_method.py
```

Output

```
[0, 2, 3, 4, 9, 11, 18]
```

The data is sorted in ascending order.

Quantum sorting methods might have the performance of O(M log M), similar to classical sorting methods. In space-bounded sorting, quantum sorting methods are better than the classical ones, as shown in Figure 7-6. Quantum algorithms help in modeling complex problems using quantum bits; when space is bounded, the methods are suitable.

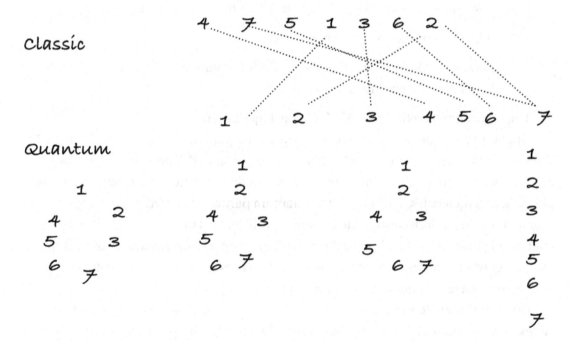

Figure 7-6. *Quantum sorting*

We will look at methods in quantum algorithms that have better sort efficiency than the classical ones. The quantum sorting method uses the quantum algorithm for ordering the elements and does this in 0.526 log M + O(M) steps compared to O(M log M), which is the classical insertion sort efficiency. The Ambainis lower bound method has the O(M) efficiency for identifying the lower bound. The algorithm finds the minimum and iterates over the list using Grover's search algorithm. This method was found by Christoph Durr and Peter Hoyer.

Grover's search algorithm can be used to find a telephone number in a phone directory. A telephone directory has the names of people, companies, and firms. The directory will be stored in a database. When the search happens online using the browser, based on the input name, the database has to be searched. The data is not sorted most of the time. It might be indexed based on some identifier chosen to improve performance. The input can be first name, middle name, and last name. Let's look at an improvised quantum sorting algorithm that helps to speed up this process. A dictionary structure can have sorted keys (quantum sorting algorithm usage) to speed up the search.

Another version of the previous quantum algorithm similar to the previous one goes through these steps:

1. The minimum element in the group is identified and stored.

2. A heap is created using the elements.

3. Iterate k through 1 to m, find the minimum from the heap, and insert the element into another heap at position k.

This algorithm has efficiency of $O(\sqrt{M}\log(1/queries))$.

Quantum entanglement happens when quantum particles interact and separate. They have the same quantum mechanical state that is shared. These entangled particles are connected in pairs. Quantum particles in the quantum system are electrons, atoms, photons, and molecules. The state of the quantum particle is related to position, spin, momentum, and polarization of the quantum particle. This phenomena of remote connected particles has challenged the relativity theory. The entangled partners impact each other when there is a change in the state. This is used in quantum cryptography and quantum teleportation–based applications.

Quantum entanglement is a unique property of quantum particle superposition states. Now let's look at the definition of superposition. A superposition state occurs when a quantum system can be present in multiple states (more than one). The superposition state can be represented as a wave function. In mathematics, the wave function is a linear combination of the quantum states with complex coefficients. In the case of entanglement, the wave functions of the twin particles cannot be separated. Mathematically, the wave function of a quantum system is not equal to the product of the particle's wave functions.

The entangled state of two quantum bits can be expressed as shown here:

$$|E1> = (1/\sqrt{2}) (|00> \pm |11>)$$
$$|E2> = (1/\sqrt{2}) (|01> \pm |10>)$$

The states of the quantum bits can be $|00>$, $|11>$, $|01>$, and $|10>$.

Getting back to sorting using quantum algorithms, the quantum entanglement principle can be used for sorting an array of M elements. The principle is based on the characteristic of quantum entanglement where if we identify one of the quantum bits, we know the other quantum bit. The algorithm goes through the following steps:

1. Input is taken as a list of M elements.

2. Elements can converted into binary representation.

3. Add 1 to the significant bit of each element.

4. Two sets are created based on the entanglement characteristic.

5. Divide the list into two sets.

6. Delete the two most significant bits.

7. Iterate from step 3 until you find a single bit. Apply a quantum single bit comparator gate.

Now let's look at some of the terms mentioned in the algorithm. The binary representation of a number is the number represented in the base-2 number system. The numerals 1 and 0 are used to represent. A zero value is 0 in binary form, a one value is equivalent to 1 in binary state, and 10 is equivalent to 2 in the binary number system. Similarly, 3, 4, and other numbers are represented in base 2 number system by adding or modifying the left and right digits, respectively.

Let's see what the definition of a significant bit is. A significant bit is the bit in the binary form with the largest value. For example, the 01000100 binary number has the most significant bit, which is 0 (which is at the start of the number from left).

Now let's look at the quantum single bit comparator definition. In classical computing, an XNOR gate is the comparator gate. The gate takes two bits as input and gives 1 if the bits are equal. Quantum comparators toggle the quantum bit in the target. This quantum bit is picked based on quantum input registers. The comparator checks the two quantum states and picks the one that is the larger value.

Quantum sorting improves the efficiency of sorting data from a database like Grover's search algorithm. Grover's search helps in searching a big datastore in the order of $O(\sqrt{M})$, whereas the classical search algorithm does this in $O(M)$. To give an example, a search in a telephone directory with 1 million records happens in 1,000 steps. The search market size is around $20 billion, and many top companies like Google and IBM are focusing on search optimization using quantum algorithms. Grover's algorithm has great potential to challenge symmetric cryptography.

Let's now look at symmetric cryptographic algorithms. An example of a symmetric cryptographic algorithm is Advanced Encryption Standard (AES). The AES algorithm is based on symmetric keys and a symmetric block cipher. The AES algorithm provides 128-, 192-, and 256-bit keys to encrypt and decrypt data. The symmetricity of the keys is in using the same key that is shared for encryption and decryption.

Now getting back to the quantum sort, let's look at Grover's search algorithm in Python.

Code

```python
import numpy as nump
import string
import hashlib
from math import sqrt, pi
from collections import OrderedDict
from statistics import mean

def GetOracleFunction(xvalue):
    return hashlib.sha256(bytes(xvalue, 'utf-8')).hexdigest()
def ExecuteGroverAlgorithm(target, objects, nvalue, rounds):
    y_pos = nump.arange(nvalue)
    amplitude = OrderedDict.fromkeys(objects, 1/sqrt(nvalue))
    for i in range(0, rounds, 2):
        for k, v in amplitude.items():
            if GetOracleFunction(k) == target:
                amplitude[k] = v * -1
        average = mean(amplitude.values())
        for k, v in amplitude.items():
            if GetOracleFunction(k) == target:
```

```python
            amplitude[k] = (2 * average) + abs(v)
            continue
        amplitude[k] = v-(2*(v-average))
    return amplitude
target_algorithm = '3'
objects_grover = ('4', '5', '3', '7','9','11','97')
number = len(objects_grover)
amplitude_grover = OrderedDict.fromkeys(objects_grover, 1/sqrt(number))

amplitude_grover[target_algorithm] = amplitude_grover[target_algorithm] * -1

average_grover = mean(amplitude_grover.values())
print("Mean is {}".format(average_grover))
for k, v in amplitude_grover.items():
    if k == target_algorithm:
        amplitude_grover[k] = (2 * average_grover) + abs(v)
        continue
    amplitude_grover[k] = v-(2*(v-average_grover))
print(amplitude_grover)
needle_value =
"2d711642b726b04401627ca9fbac32f5c8530fb1903cc4db02258717921a4881"
haystack_value = string.ascii_lowercase
#print(haystack_value)
num = len(haystack_value)
num_rounds = int((pi / 4) * sqrt(num))
print("number of rounds are {}".format(num_rounds))
amplitude_grover = ExecuteGroverAlgorithm(needle_value, haystack_value,
num, num_rounds)
print(amplitude_grover)
```

Command

```
pip3 install numpy
pip3 install hashlib
python3 quantum_sort.py
```

Output

```
Mean is 0.2699746235780194OrderedDict([('4', 0.16198477414681167),
('5', 0.16198477414681167), ('3', 0.9179137201652661), ('7',
0.16198477414681167), ('9', 0.16198477414681167), ('11',
0.16198477414681167), ('97', 0.16198477414681167)])


number of rounds are 4
OrderedDict([('a', 0.11024279785874247), ('b', 0.11024279785874247),
('c', 0.11024279785874247), ('d', 0.11024279785874247), ('e',
0.11024279785874247), ('f', 0.11024279785874247), ('g',
0.11024279785874247), ('h', 0.11024279785874247), ('i',
0.11024279785874247), ('j', 0.11024279785874247), ('k',
0.11024279785874247), ('l', 0.11024279785874247), ('m',
0.11024279785874247), ('n', 0.11024279785874247), ('o',
0.11024279785874247), ('p', 0.11024279785874247), ('q',
0.11024279785874247), ('r', 0.11024279785874247), ('s',
0.11024279785874247), ('t', 0.11024279785874247), ('u',
0.11024279785874247), ('v', 0.11024279785874247), ('w',
0.11024279785874247), ('x', 0.8343639122151143), ('y',0.11024279785874247),
('z', 0.11024279785874247)])
```

Tip   The oracle function is a black box that is part of an algorithm. The oracle function will be used as an input to another algorithm.

Let's now look at quantum eigen solvers.

Quantum Eigen Solvers

Quantum eigen solvers help in identifying eigenvalues of a matrix G. The solver calculates the upper limit of the lowest eigenvalue of a specific Hamiltonian. This method has better performance over iterative quantum phase estimation and quantum phase estimation techniques. Quantum phase estimation techniques are implemented to calculate the ground state energy of a molecule.

Tip   The Hamiltonian matrix is used to present the feasible energies of the system. It represents different states of the system.

The system in context can exist in multiple states, and every state has a specified energy value. The eigenvalues are the state values of the system, and they represent the energies associated with the system state.

Now let's look at quantum semidefinite programming.

Quantum Semidefinite Programming

Quantum semidefinite programming is related to convex optimization. Convex optimization is based on linear functions and constraints. Constraints are mathematical equalities or inequalities. Quantum semidefinite programming was discussed in detail in Chapter 6 with code examples.

Summary

In this chapter, we looked at the techniques related to quantum algorithms such as quantum least squares fitting and quantum sort. The concepts of quantum eigen solvers were discussed in detail.

## CHAPTER 8

Quantum Neural Network Algorithms

Introduction

"Spurred on by both the science and science fiction of our time, my genera-tion of researchers and engineers grew up to ask what if? and what's next? We went on to pursue new disciplines like computer vision, artificial intel-ligence, real-time speech translation, machine learning, and quantum computing."

—Elizabeth Bear

This chapter covers the topic of quantum neural networks. You will see how quantum neural networks are implemented in real life. Code examples are presented for the quantum neural network topics related to quantum artificial neural networks (ANNs), quantum associative memory, quantum dots, quantum generative adversarial networks (GANs) , and quantum random access memory.

Initial Setup

You need to set up Python 3.5 to run the code samples in this chapter. You can download it from https://www.python.org/downloads/release/python-350/.

© Bhagvan Kommadi 2020
B. Kommadi, *Quantum Computing Solutions*, https://doi.org/10.1007/978-1-4842-6516-1_8

Quantum ANN

In this section, we will look at quantum artificial neural networks. Quantum neural networks are the quantum equivalent of neural networks. They are based on quantum mechanics principles. Quantum neural networks are based on quantum information processing and brain quantum effects. Quantum artificial neural networks are used for pattern classification algorithms.

Tip   Pattern classification is related to identifying the right group for an input value given the different possible groups.

Let's start looking at classical neural networks.

Classical Neural Networks

A neural network has layers that are input, output, and hidden types. The network changes the input depending on the usage of the output layer. The concept of neural networks is based on the human brain. Neural networks have been part of AI since the 1940s. The reason for the addition was the backpropagation method, which helps in changing the hidden layers when the output's expected value is not equal to the input's outcome.

For example, a neural network that is designed to identify cricket balls can make a mistake and instead identify a tennis ball. Deep learning neural networks are multilayer advanced neural networks. Deep learning neural networks retrieve multiple features to identify the objects in context.

A deep learning neural network is like a conveyor belt that takes a crude iron rod and transforms it to a specific designed part. Each stage in the deep learning neural network has the capability to identify features and retrieve them. Artificial neural networks are based on mathematical models. They consist of elements that are interconnected that are referred to as *neurons*. The neurons have information that is equal to the weighted linkage. The neuron signals that are the input values pass through neural connections, and the connections are weighted. A deep learning neural network has capacity to learn, modify, and normalize the input data by using the appropriate weights.

Quantum Analogies to Neural Networks

After looking at the classical neural networks, we can now look at the quantum equivalent of neural networks. Classical ANNs have computational power because of their capability to process big data in parallel mode. Quantum neural networks can handle similar big data because of quantum parallelism based on the quantum mechanics principles. A quantum register that has M quantum bits has different configurations that can be handled by the quantum computer.

Quantum entanglement and superposition are the key principles that help in measuring the quantum bit states. A quantum neural network consists of input, hidden, and output neurons that are connected. An example of a quantum neural network is the Hopfield model. The Hopfield model became part of ANNs in 1982. John Hopfield discovered the symmetrical bonds and quantum spin glasses in neural networks.

Tip   Hopfield networks are applied in image processing and analysis, as well as medical image analysis and restoration.

Neurocomputational methods are based on the nonlinear techniques of data processing. Quantum computational methods are evolutionary as they are based on the evolutionary operators of quantum mechanics. Copenhagen interpretation states that evolutionary behavior is blended with nonevolutionary behavior in quantum systems. A pattern recall is not an evolutionary operator. It is a nonunitary process. Feynman interpretation of quantum mechanics uses path integrals. Elizabeth Behrman and Goertzel proposed a temporal model for a quantum neural network. See Figure 8-1.

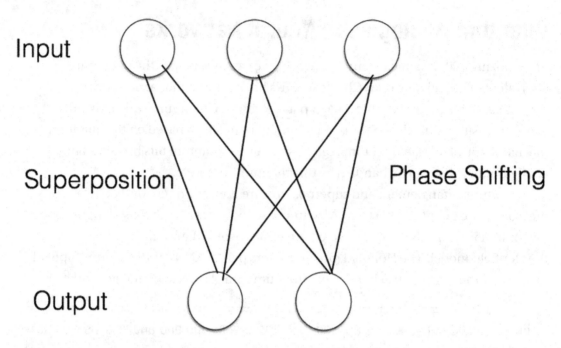

Figure 8-1. *Quantum neural network*

A quantum neural network uses a quantum dot molecule and optical photons in the model. The temporal model of a QNN is modeled by the mathematical equations of the virtual neurons. The number of virtual neurons in a QNN is determined by the temporal discretization parameters. A new model was created by Wichita State University that is based on the spatial model of QNN. The spatial model uses the array of quantum dot molecules.

Ron Chrisley, who was based in the University of Sussex, has proposed a different approach related to the network weights that are dependent on the position of silts. A quantum neural network that is based on neurocomputing is used for pattern recognition in real life.

Code Sample

```
import pennylane as qnnL
from pennylane import numpy as numcal
from pennylane.optimize import AdamOptimizer as AdamO

try:
    device = qnnL.device('strawberryfields.fock', wires=1, cutoff_dim=10)
```

```python
except:
    print("please install the package strawberryfields")

def GetQNNLayer(varr):
    """ One layer of the quantum neural network
    """
    qnnL.Rotation(varr[0], wires=0)
    qnnL.Squeezing(varr[1], 0., wires=0)
    qnnL.Rotation(varr[2], wires=0)

    qnnL.Displacement(varr[3], 0., wires=0)

    qnnL.Kerr(varr[4], wires=0)

@qnnL.qnode(device)
def GetQNN(vars, xcoor=None):
    """ Returns The quantum neural network
    """
    qnnL.Displacement(xcoor, 0., wires=0)

    for var in vars:
    GetQNNLayer(var)

    return qnnL.expval.X(0)

def GetQNNSquareLoss(labelValues, predictionValues):
    """ Get the Square loss function
    """
    lossValue = 0
    for label, prediction in zip(labelValues, predictionValues):
    lossValue = lossValue + (label - prediction) ** 2
    lossValue = lossValue / len(labelValues)

    return lossValue

def GetQNNCost(variables, featureValues, labelValues):
    """ Minimizing Cost function
    """
```

```python
    predictions = [GetQNN(variables, xcoor=x) for x in featureValues]

    return GetQNNSquareLoss(labelValues, predictions)

sincurvevalues = numcal.loadtxt("sin_curve_values.txt")
xcoordinate= sincurvevalues[:, 0]
ycoordinate = sincurvevalues[:, 1]

numcal.random.seed(0)
layer_recount = 4
variable_initial = 0.05 * numcal.random.randn(layer_recount, 5)

optimizer = AdamO(0.01, beta1=0.9, beta2=0.999)

variable = variable_initial

for iteration in range(500):
    var_cost = optimizer.step(lambda v: GetQNNCost(v, xcoordinate,
    ycoordinate), variable)

    print("Iteration value: {:5d} | Cost is: {:0.7f} ".format(iteration +
    1, GetQNNCost(var_cost, xcoordinate, ycoordinate)))
```

Command

```
pip3 install numpy

pip3 install pennylane

python3 qnn.py
```

Output

```
Iteration value:      1 | Cost is: 0.2689702
Iteration value:      2 | Cost is: 0.2768397
Iteration value:      3 | Cost is: 0.2801266
Iteration value:      4 | Cost is: 0.2819382
Iteration value:      5 | Cost is: 0.2830613
Iteration value:      6 | Cost is: 0.2837995
```

```
Iteration value:      7 | Cost is: 0.2842974
Iteration value:      8 | Cost is: 0.2846337
Iteration value:      9 | Cost is: 0.2848555
Iteration value:     10 | Cost is: 0.2849926
..........................................
..........................................

..........................................
Iteration value:    494 | Cost is: 0.2406537
Iteration value:    495 | Cost is: 0.2406112
Iteration value:    496 | Cost is: 0.2405687
Iteration value:    497 | Cost is: 0.2405263
Iteration value:    498 | Cost is: 0.2404840
Iteration value:    499 | Cost is: 0.2404417
Iteration value:    500 | Cost is: 0.2403996
```

Quantum Associative Memory

Quantum associative memory is based on a quantum neurocomputing approach. Memorization and recall are the parts of pattern association. The patterns are persisted in the memorization step. The recall step consists of pattern association and completion.

Memorization is a quantum algorithm for creating a quantum state using "m" quantum bits to model "q" patterns. This algorithm uses the polynomial number of basic operations using 1, 2, and 3 quantum bits. The algorithm uses controlled-NOT and Fredkin quantum gates for mathematical operations. This algorithm executes in the order of mq steps to model "q" patterns using "m" quantum neurons.

The recall step in the quantum associative memory is based on Grover's search algorithm. The search algorithm finds the item in the database of m items in the order of $\sqrt{m}$ time. Memorization and recall algorithms are mixed in to create a quantum associative memory process.

The quantum associative memory algorithm was created by Dan Ventura and Tony Martinez. The algorithm is related to associative memory simulation on quantum circuit–based computers. The quantum bits are in superposition state in the quantum associative memory. Grover's algorithm can be modified using the quantum associative memory algorithm for search via the memory state based on the search input.

Quantum associative memory is used extensively in the field of neural networks. This type of memory has storage exponential to the neuron's count in the neural network. This memory model is based on the human memory that has neural network–based processing in the brain.

Tip   The quantum associative memory algorithm is based on a polynomial-time algorithm that creates a quantum state using a quantum bit set based on the training data.

Let's now look at the concept called *quantum dots.*

Quantum Dots

Quantum dots are particles that are nanometers in size. These particles emit light when they are exposed to light. The wavelength of the particle's emission can be modified by the size, shape, and material of the quantum dots. Blue-green spectral regions are created by the quantum dots with a size of 2 or 3 nanometers. Five to six nanometers-sized quantum dots emit light related to the orange, red, and infrared spectral regions. Quantum dot size is related to the lifetime of emissions. The larger the quantum dot, the longer the lifetime.

Nanoparticles are larger in size than quantum dots. Nanoparticles are 8 to 100 nanometers. Quantum dot devices are based on the quantum dots. In a quantum dot device, the quantum dots are combined with leads using tunneling. Different methods of first order are used for computing the currents of the stationary state particles. The methods used by Pauli, Lindblad, Red-field, and Von Neumann are used to compute the effect of Coulomb blockade. There are second-order methods such as Von Neumann's -2vN that can be used for co-tunneling, pair tunneling, and broadening reasons.

Tip   Quantum dots help to manufacture the fluorescence and FRET-based biosensors. The biosensors are used for identifying biomolecules such as acids, enzymes, sugars, antibodies, and antigens.

Now let's look at the quantum random access memory concept.

Quantum Random Access Memory

Quantum random access memory is based on m quantum bits to create an address of 2^m distinct memory cells. This memory model is used for superposition of m memory cells. This model is based on $O(\log 2^m)$ switches opposed to quantum RAM traditional design. This model helps to address using lesser power. This is achieved using the entanglement principle.

QRAM is the basis for quantum computers. Address and output registers based on quantum bits help in memory management in QRAM. Memory access algorithms are based on coherent quantum superposition. QRAM is used for quantum searching, collision identification, element uniqueness, coherent routing of signals, and encrypted data searches.

Let's look at quantum generative adversarial network algorithms.

Quantum GAN

In classical machine learning, generative adversarial networks are a tool. A generator is used to create statistics for data. It tries to mimic the current data set. A discriminator differentiates between the real and fake data. The learning process for a generator and a discriminator converges. The convergence happens when the generator generates the statistics. At the same time, the discriminator cannot differentiate between the real and generated data. See Figure 8-2.

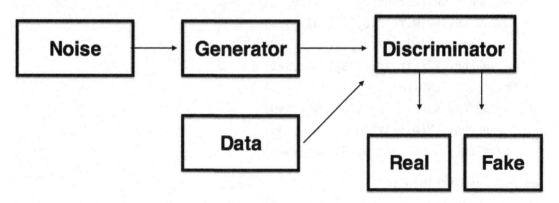

Figure 8-2. *QGAN*

The quantum generative adversarial networks (QGANs) are generative adversarial networks using quantum processors. Seth Lloyd and Christian Weed Brook first devised the adversarial learning algorithm. This algorithm runs on a quantum computer that has a quantum register with qubits. Qubits exist in the 0 and 1 states. Quantum computers process vectors in very high-dimensional vector spaces. These vectors have information to be processed.

Let's look at an implementation of a quantum generative adversarial network. The implementation is based on two quantum circuits: the generator and the other discriminator.

Code Sample

```
import pennylane as qganPenny
from pennylane import numpy as numcal
from pennylane.optimize import GradientDescentOptimizer as GDO

device = qganPenny.device('default.qubit', wires=3)

def GetQGANReal(phi, theta, omega):
    qganPenny.Rot(phi, theta, omega, wires=0)

def GetQGANGenerator(wireArray):
    qganPenny.RX(wireArray[0], wires=0)
    qganPenny.RX(wireArray[1], wires=1)
    qganPenny.RY(wireArray[2], wires=0)
    qganPenny.RY(wireArray[3], wires=1)
    qganPenny.RZ(wireArray[4], wires=0)
    qganPenny.RZ(wireArray[5], wires=1)
    qganPenny.CNOT(wires=[0,1])
    qganPenny.RX(wireArray[6], wires=0)
    qganPenny.RY(wireArray[7], wires=0)
    qganPenny.RZ(wireArray[8], wires=0)

def GetQGANDiscriminator(wireArray):
    qganPenny.RX(wireArray[0], wires=0)
    qganPenny.RX(wireArray[1], wires=2)
    qganPenny.RY(wireArray[2], wires=0)
```

```
    qganPenny.RY(wireArray[3], wires=2)
    qganPenny.RZ(wireArray[4], wires=0)
    qganPenny.RZ(wireArray[5], wires=2)
    qganPenny.CNOT(wires=[1,2])
    qganPenny.RX(wireArray[6], wires=2)
    qganPenny.RY(wireArray[7], wires=2)
    qganPenny.RZ(wireArray[8], wires=2)

@qganPenny.qnode(device)
def GetQGANRealDiscCircuit(phi, theta, omega, discWeights):
    GetQGANReal(phi, theta, omega)
    GetQGANDiscriminator(discWeights)
    return qganPenny.expval.PauliZ(2)

@qganPenny.qnode(device)
def GetQGANDiscCircuit(genWeights, discWeights):
    GetQGANGenerator(genWeights)
    GetQGANDiscriminator(discWeights)
    return qganPenny.expval.PauliZ(2)

def GetQGANRealTrue(discWeights):
    trueDiscriminatorOutput = GetQGANRealDiscCircuit(phi, theta, omega,
    discWeights)
    probabilityRealTrue = (trueDiscriminatorOutput + 1) / 2
    return probabilityRealTrue

def GetQGANFakeTrue(genWeights, discWeights):
    fakeDiscriminatorOutput = GetQGANDiscCircuit(genWeights, discWeights)
    probabilityFakeTrue = (fakeDiscriminatorOutput + 1) / 2
    return probabilityFakeTrue

def GetQGANDiscriminatorCost(discWeights):
    cost = GetQGANFakeTrue(genWeights, discWeights) -
    GetQGANRealTrue(discWeights)
    return cost

def GetQGANGeneratorCost(genWeights):
    return -GetQGANFakeTrue(genWeights, discWeights)
```

```
phi = numcal.pi / 6
theta = numcal.pi / 2
omega = numcal.pi / 7

numcal.random.seed(0)
epsValue = 1e-2
genWeights = numcal.array([numcal.pi] + [0] * 8) + numcal.random.
normal(scale=epsValue, size=[9])
discWeights = numcal.random.normal(size=[9])

gdo = GDO(0.1)

print("Training the discriminator ")
for iteration in range(50):
    discriminator_weights = gdo.step(GetQGANDiscriminatorCost,
    discWeights)
    discriminator_cost = GetQGANDiscriminatorCost(discriminator_weights)
    if iteration % 5 == 0:
        print("Iteration {}: discriminator cost = {}".format(iteration+1,
        discriminator_cost))

print(" discriminator - real true: ", GetQGANRealTrue(discWeights))
print("discriminator - fake true: ", GetQGANFakeTrue(genWeights,
discWeights))

print("Training the generator.")

for iteration in range(200):
    generator_weights = gdo.step(GetQGANGeneratorCost, genWeights)
    generator_cost = -GetQGANGeneratorCost(generator_weights)
    if iteration % 5 == 0:
        print("Iteration num {}: generator cost is = {}".format(iteration,
        generator_cost))

print("discriminator - real true: ", GetQGANRealTrue(discWeights))
print("discriminator -  fake true: ", GetQGANFakeTrue(genWeights,
discWeights))

print("Cost is: ", GetQGANDiscriminatorCost(discriminator_weights))
```

Command

```
pip3 install numpy

pip3 install pennylane

python3 qgan.py
```

Output

```
Training the discriminator
Iteration num 1: discriminator cost is = -0.10942017805789161
Iteration num 6: discriminator cost is = -0.10942017805789161
Iteration num 11: discriminator cost is = -0.10942017805789161
Iteration num 16: discriminator cost is = -0.10942017805789161
Iteration num 21: discriminator cost is = -0.10942017805789161
Iteration num 26: discriminator cost is = -0.10942017805789161
Iteration num 31: discriminator cost is = -0.10942017805789161
Iteration num 36: discriminator cost is = -0.10942017805789161
Iteration num 41: discriminator cost is = -0.10942017805789161
Iteration num 46: discriminator cost is = -0.10942017805789161
 discriminator - real true:  0.1899776355995153
discriminator - fake true:  0.13399334520305328
Training the generator.
Iteration num 0: generator cost is = 0.1339934177809623
Iteration num 5: generator cost is = 0.1339934177809623
Iteration num 10: generator cost is = 0.1339934177809623
Iteration num 15: generator cost is = 0.1339934177809623
Iteration num 20: generator cost is = 0.1339934177809623
Iteration num 25: generator cost is = 0.1339934177809623
Iteration num 30: generator cost is = 0.1339934177809623
Iteration num 35: generator cost is = 0.1339934177809623
Iteration num 40: generator cost is = 0.1339934177809623
Iteration num 45: generator cost is = 0.1339934177809623
Iteration num 50: generator cost is = 0.1339934177809623
Iteration num 55: generator cost is = 0.1339934177809623
Iteration num 60: generator cost is = 0.1339934177809623
```

```
Iteration num 65: generator cost is = 0.13399341778096623
Iteration num 70: generator cost is = 0.13399341778096623
Iteration num 75: generator cost is = 0.13399341778096623
Iteration num 80: generator cost is = 0.13399341778096623
Iteration num 85: generator cost is = 0.13399341778096623
Iteration num 90: generator cost is = 0.13399341778096623
Iteration num 95: generator cost is = 0.13399341778096623
Iteration num 100: generator cost is = 0.13399341778096623
Iteration num 105: generator cost is = 0.13399341778096623
Iteration num 110: generator cost is = 0.13399341778096623
Iteration num 115: generator cost is = 0.13399341778096623
Iteration num 120: generator cost is = 0.13399341778096623
Iteration num 125: generator cost is = 0.13399341778096623
Iteration num 130: generator cost is = 0.13399341778096623
Iteration num 135: generator cost is = 0.13399341778096623
Iteration num 140: generator cost is = 0.13399341778096623
Iteration num 145: generator cost is = 0.13399341778096623
Iteration num 150: generator cost is = 0.13399341778096623
Iteration num 155: generator cost is = 0.13399341778096623
Iteration num 160: generator cost is = 0.13399341778096623
Iteration num 165: generator cost is = 0.13399341778096623
Iteration num 170: generator cost is = 0.13399341778096623
Iteration num 175: generator cost is = 0.13399341778096623
Iteration num 180: generator cost is = 0.13399341778096623
Iteration num 185: generator cost is = 0.13399341778096623
Iteration num 190: generator cost is = 0.13399341778096623
Iteration num 195: generator cost is = 0.13399341778096623
discriminator - real true:  0.18997763559951530
discriminator -  fake true:  0.13399334520305328
Cost is:  -0.10942017805789161
```

Summary

In this chapter, we looked at the techniques related to quantum neural network algorithms such as quantum neural networks and quantum generative adversarial networks. The chapter discussed in detail the concepts of quantum dots, quantum associative memory, and quantum random access memory.

## CHAPTER 9

Quantum Classification Algorithms

Introduction

The point is no longer that quantum mechanics is an extraordinarily (and for Einstein, unacceptably) peculiar theory, but that the world is an extraordinarily peculiar place.

—N. David Mermin

This chapter covers the topic of quantum classification algorithms. You will see how quantum classification algorithms are implemented in real life. Code examples are presented for different algorithms such as quantum classifiers, support vector machines, and sparse support vector machines.

Initial Setup

You need to set up Python 3.5 to run the code samples in this chapter. You can download it from `https://www.python.org/downloads/release/python-350/`.

Classifiers

Patterns are identified by using machine learning algorithms in classical computing. Quantum machine learning techniques help to convert machine learning methods to algorithms. Algorithms are the basic units of quantum computing. New research is going on in the quantum computing area; specifically, quantum computing is evolving by

© Bhagvan Kommadi 2020
B. Kommadi, *Quantum Computing Solutions*, https://doi.org/10.1007/978-1-4842-6516-1_9

adding data mining to it. Pattern identification can be done using quantum computing algorithms. Supervised learning methods operate on labeled input data sets and feature vectors. These methods help in identifying the labels for feature vectors. See Figure 9-1.

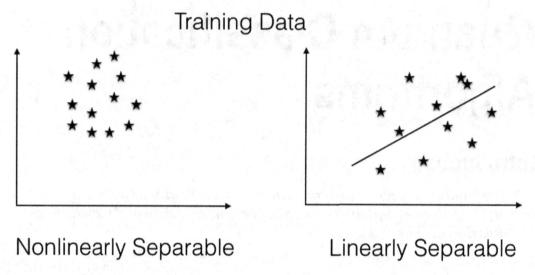

Figure 9-1. *Linearly separable*

The data set, which is linearly separable, splits the input set by a plane, line, or hyperplane. The points of one set are in the first half-space, and the second set are in the other space.

Quantum machine learning methods are based on supervised and unsupervised learning. Amplitude modification methods are used in the k-nearest neighbor and clustering quantum algorithms. Quantum matrix computations are used for quantum kernel methods. Quantum kernel methods support vector machines and Gaussian processes. Supervised quantum machine learning is the basis for variational quantum classifier methods.

Let's look at quantum classifiers in detail.

Quantum Classifiers

Quantum classifiers consist of components such as an encoder and a processor (see Figure 9-2). The encoder modifies the data vector to a quantum state. A classical preprocessor changes the input data to a quantum circuit. The quantum circuit is initialized to a zero quantum state. The processor component retrieves the information from the encoded quantum state. The quantum state transformer changes the quantum circuit's encoded quantum state to information retrieved by measurement and post-processing.

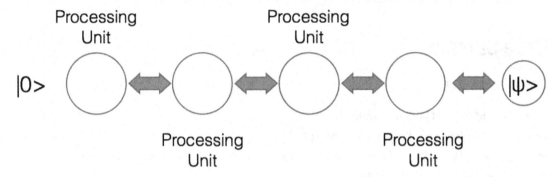

Figure 9-2. Quantum classifier

We're using the QClassify framework to demonstrate quantum classifiers. QClassifiers are variational quantum classifiers. This package has classes for classifier tasks. Classifier tasks are based on quantum gates. A quantum simulator on the cloud is used for simulating quantum gates on the quantum circuit.

A quantum circuit has a single Hadamard gate and two single quantum bit measure gates. Measure gates operate on an ancilla and designated quantum bit. Quantum methods modify the data to encoded amplified data format. A QClassifier is used to build a concrete circuit. The QClassify object has the component information. This package is used for customization based on the context of the problem. Quantum binary classifiers are based on quantum kernel methods.

Tip   Kernels are mathematical functions used to calculate the distance between any two points on a graph or in a word.

Let's look at an example related to the binary classification of a set of two-dimensional points. This example is modeled as a quantum circuit of 2 quantum bits. The processor has multiple options such as layer_xyz, measure_top, and prob_one. Quantum bits can have 0 or 1 values or be in a superposition state of 0 and 1. The encoder options are also set. Classifiers are created with selected quantum bits and classifier options.

An input vector has [1,1] matrix values. Parameters have 3.067 and 3.3311 values. The quantum circuit is created with input_vector and parameters. The program is the output of the quantum circuit.

Let's look at the code implementation of a quantum classifier.

Code Sample

```
import qclassify

from qclassify.preprocessing import *
from qclassify.encoding_circ import *
from qclassify.qclassifier import QClassifier as QuantumClassifier
from qclassify.proc_circ import *
from qclassify.postprocessing import *

encoding_opts={
                'preprocessing':id_func,
                'encoding_circ':x_product,
        }

process_opts={
                'proc_circ':layer_xz,
                'postprocessing':{
                        'quantum':measure_top,
                        'classical':prob_one,
                }
        }
quantum_bits = [0, 1]
classify_opts = {
                'encoder_options':encoding_opts,
                'proc_options':process_opts,
        }
qCircuit = QuantumClassifier(quantum_bits, classify_opts)

circuit_values = [1, 1]
params = [2.0672044712460114,          1.3311348339721203]
```

```
quantum_circuit_prog = qCircuit.circuit(circuit_values, params)

print(quantum_circuit_prog)
```

Command

```
pip3 install qclassify

python3 quantum_classifier.py
```

Output

```
RX(1) 0
RX(1) 1
CZ 0 1
RX(3.0672044712460114) 0
RX(3.3311348339721203) 1
DECLARE ro BIT[1]
MEASURE 0 ro[0]
```

Now, let's look at the variational quantum classifier in detail.

Variational Quantum Classifier

Quantum algorithms operate on quantum bits and gates. Error-proof gates and quantum bits are necessary for the algorithms. Variational quantum circuits are used in machine learning. In these circuits, gate parameters are trained using learning methods. The input vectors are the features of the training model. They are equivalent to the quantum system amplitudes. A variational quantum circuit has 1 and 2 quantum bit gates. It has a measurement gate with one quantum bit to classify the inputs. The training method is based on the quantum methods. See Figure 9-3.

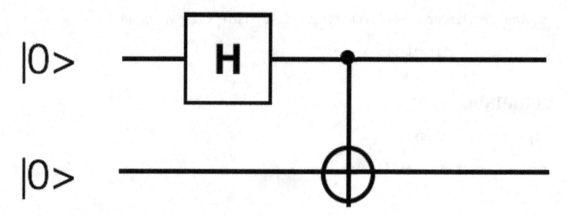

Figure 9-3. *Quantum circuit*

Let's look at the iris flower problem where patterns are identified in different species of the flower. An iris has three different types: setosa, virginica, and versicolor. The patterns identified in these flowers are length and width of petals and sepals. The size is measured in centimeters. The data is split into training and validation data sets. Each data set has the petal length, petal width, sepal width, and sepal length of different classes of flowers. See Figure 9-4.

Figure 9-4. *Iris flower (source: https://pixabay.com/photos/iris-flower-flora-blue-blossom-2339883/)*

The classifying solution iterates over the data set, and accuracy is achieved when using the quantum machine learning model and data. The variational quantum classifier algorithm is based on supervised learning.

We can use the PennyLane framework for quantum computation and optimization (see Figure 9-5). The package has a quantum simulator to execute quantum circuits. Variational quantum circuits are simulated on the quantum circuits with parameters that are trained. PennyLane has features related to gates, state operations, and measurement gates. It can handle multiple types of measurement gates. The package has templates based on different variational quantum circuits used for creating and training machine learning models. Embeddings, layers, subroutines, and state preparations are multiple types of quantum circuits. These circuits are called *ansatz circuits*.

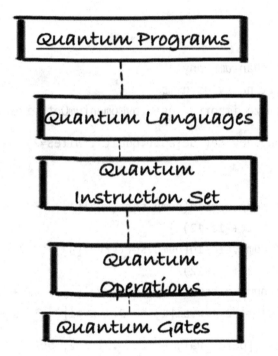

Figure 9-5. *PennyLane framework*

Quantum optimizers are used to modify the quantum machine learning models. The package has multiple types of optimizer such as gradient descent optimizer, Nesterov momentum–based gradient descent optimizer, adaptive learning–based optimizer, and root mean squared propagation optimizer.

The variational quantum classifier algorithm is based on supervised learning. A quantum circuit that has 1 and 2 quantum bit gates is modeled and simulated on the quantum device. The code sample in the next section shows the implementation for calculating the rotation angles for the quantum state. The class has methods to encode a vector 4D based on 2 quantum bit amplitudes.

Methods in the class have features for creating a variational classifier, creating a quantum circuit based on the variational classifier, calculating the loss in prediction, predicting the accuracy, and calculating the cost. The data set has the coordinates x, y, and has pad values. The code is executed for 80 steps for optimization. The accuracy and the cost function are printed for every iteration.

Let's look at the code sample for a variational quantum classifier.

Code Sample

```
import pennylane as quantumPenny
from pennylane import numpy as nump
from pennylane.optimize import  NesterovMomentumOptimizer as NMO

device = quantumPenny.device('default.qubit', wires=2)

def CalcAngles(xval):

    beta0_val = 2 * nump.arcsin(nump.sqrt(xval[1]) ** 2 / nump.sqrt(xval[0]
    ** 2 + xval[1] ** 2 + 1e-12) )
    beta1_val = 2 * nump.arcsin(nump.sqrt(xval[3]) ** 2 / nump.sqrt(xval[2]
    ** 2 + xval[3] ** 2 + 1e-12) )
    beta2_val = 2 * nump.arcsin(nump.sqrt(xval[2] ** 2 + xval[3] ** 2) /
    nump.sqrt(xval[0] ** 2 + xval[1] ** 2 + xval[2] ** 2 + xval[3] ** 2))

    return nump.array([beta2_val, -beta1_val / 2, beta1_val / 2,
    -beta0_val / 2, beta0_val / 2])

def CalcState(arr):

    quantumPenny.RY(arr[0], wires=0)

    quantumPenny.CNOT(wires=[0, 1])
    quantumPenny.RY(arr[1], wires=1)
```

```python
    quantumPenny.CNOT(wires=[0, 1])
    quantumPenny.RY(arr[2], wires=1)

    quantumPenny.PauliX(wires=0)
    quantumPenny.CNOT(wires=[0, 1])
    quantumPenny.RY(arr[3], wires=1)
    quantumPenny.CNOT(wires=[0, 1])
    quantumPenny.RY(arr[4], wires=1)
    quantumPenny.PauliX(wires=0)

def CalcLayer(Warr):
    quantumPenny.Rot(Warr[0, 0], Warr[0, 1], Warr[0, 2], wires=0)
    quantumPenny.Rot(Warr[1, 0], Warr[1, 1], Warr[1, 2], wires=1)

    quantumPenny.CNOT(wires=[0, 1])

@quantumPenny.qnode(device)
def FindCircuit(weight, angle=None):

    CalcState(angle)

    for W in weight:
        CalcLayer(W)

    return quantumPenny.expval(quantumPenny.PauliZ(0))

def GetVariationalClassifier(var, angle=None):

    weight = var[0]
    bias_val = var[1]

    return FindCircuit(weight, angle=angle) + bias_val

def CalcSquareLoss(label, prediction):
    loss_val = 0
    for l, p in zip(label, prediction):
        loss_val = loss_val + (l - p) ** 2
    loss_val = loss_val / len(label)

    return loss_val
```

```python
def CalcAccuracy(label, prediction):

    loss_val = 0
    for l, p in zip(label, prediction):
        if abs(l - p) < 1e-5:
            loss_val = loss_val + 1
    loss_val = loss_val / len(label)

    return loss_val

def CalcCost(weight, feature, label):

    prediction = [GetVariationalClassifier(weight, angle=f) for f in
    feature]

    return CalcSquareLoss(label, prediction)

data_iris_flowers = nump.loadtxt("iris_flower_data.txt")
XVal = data_iris_flowers[:, 0:2]

pad = 0.3 * nump.ones((len(XVal), 1))
XPad = nump.c_[nump.c_[XVal, pad], nump.zeros((len(XVal), 1)) ]

norm = nump.sqrt(nump.sum(XPad ** 2, -1))
X_norm = (XPad.T / norm).T

ftrs = nump.array([CalcAngles(x) for x in X_norm])

YVal = data_iris_flowers[:, -1]

nump.random.seed(0)
num_data = len(YVal)
num_train = int(0.75 * num_data)
index = nump.random.permutation(range(num_data))
feats_train = ftrs[index[:num_train]]
YValT = YVal[index[:num_train]]
ftrs_val = ftrs[index[num_train:]]
Y_val = YVal[index[num_train:]]

quantumbits = 2
layers = 6
```

```
varinit = (0.01 * nump.random.randn(layers, quantumbits, 3), 0.0)

opt = NMO(0.01)
batsize = 5

var = varinit
for it in range(50):

    bat_index = nump.random.randint(0, num_train, (batsize, ))
    features_train_batch = feats_train[bat_index]
    Ytrainbat = YValT[bat_index]
    var = opt.step(lambda v: CalcCost(v, features_train_batch,
    Ytrainbat), var)

    predict_train = [nump.sign(GetVariationalClassifier(var, angle=f)) for
    f in feats_train]
    predict_val = [nump.sign(GetVariationalClassifier(var, angle=f)) for f
    in ftrs_val]

    accuracy_train = CalcAccuracy(YValT, predict_train)
    accuracy_val = CalcAccuracy(Y_val, predict_val)

    print("it value: {:5d} | Cost Features: {:0.7f} | Accuracy training:
    {:0.7f} | Accuracy for validation: {:0.7f} "
        "".format(it+1, CalcCost(var, ftrs, YVal), accuracy_train,
        accuracy_val))
```

Command

```
pip3 install pennylane
python3 variational_classifier.py
```

Output

```
it value:    1 | Cost Features: 1.4490948 | Accuracy training: 0.4933333 |
Accuracy for validation: 0.5600000
it value:    2 | Cost Features: 1.3309953 | Accuracy training: 0.4933333 |
Accuracy for validation: 0.5600000
```

```
it value:      3 | Cost Features: 1.1582178 | Accuracy training: 0.4533333 |
Accuracy for validation: 0.5600000
it value:      4 | Cost Features: 0.9795035 | Accuracy training: 0.4800000 |
Accuracy for validation: 0.5600000
it value:      5 | Cost Features: 0.8857893 | Accuracy training: 0.6400000 |
Accuracy for validation: 0.7600000
it value:      6 | Cost Features: 0.8587935 | Accuracy training: 0.7066667 |
Accuracy for validation: 0.7600000
it value:      7 | Cost Features: 0.8496204 | Accuracy training: 0.7200000 |
Accuracy for validation: 0.6800000
it value:      8 | Cost Features: 0.8200972 | Accuracy training: 0.7333333 |
Accuracy for validation: 0.6800000
it value:      9 | Cost Features: 0.8027511 | Accuracy training: 0.7466667 |
Accuracy for validation: 0.6800000
it value:     10 | Cost Features: 0.7695152 | Accuracy training: 0.8000000 |
Accuracy for validation: 0.7600000
.............................................................................
.............................................................................
it value:     47 | Cost Features: 0.5413240 | Accuracy training: 0.9466667 |
Accuracy for validation: 1.0000000
it value:     48 | Cost Features: 0.5239643 | Accuracy training: 0.9466667 |
Accuracy for validation: 1.0000000
it value:     49 | Cost Features: 0.5100842 | Accuracy training: 0.9466667 |
Accuracy for validation: 1.0000000
it value:     50 | Cost Features: 0.5006861 | Accuracy training: 0.9466667 |
Accuracy for validation: 1.0000000
```

The training data and test data are about 75% of the data set. The validation set is about 25% of the data set values. The training data is used to fit the parameters of the model, and the test data is used to identify the model's performance. The validation data is used to fine-tune the machine learning model.

Loss and accuracy are measured on the training and validation data sets. Overfitting of the model occurs when the validation loss increases versus the training loss. The goal is to have the validation accuracy high compared to the loss. Representative data sets need to be chosen to avoid data source bias. Bias can be estimated based on the statistical expected value and the real value from the data population. The model error that is observed needs to be calculated based on a metric mean squared error. The mean squared error needs to be minimized.

Classical SVM

A classical support vector machine is based on supervised machine learning (SVM). SVM is trained using the data, and it is used to forecast whether the input belongs to a specific class. The training data has the set of classes that SVM forecasts. Support vectors are used in SVM to maximize the distance between the set of classes. See Figure 9-6.

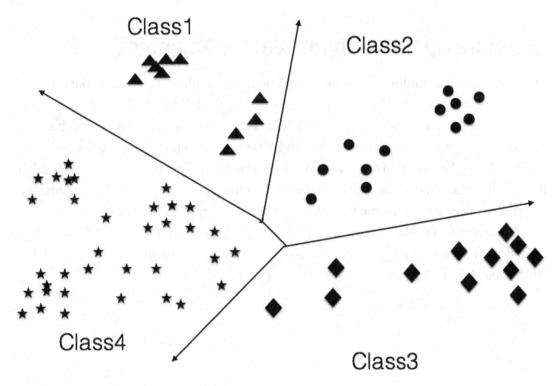

Figure 9-6. *Support vector machine*

The quantum equivalent of SVM uses a feature map to present the quantum circuit information. The kernel concept is used in classification methods. A feature map modified the features. Features are measurable attributes of the subject in context. Classification is about finding the distance between the classes for the input information.

The quantum support vector machine is based on a quantum processor for solutions. It is based on supervised learning. Multiclass classifiers are built on quantum support vector machines. A quantum processor has QRAM for handling quantum bits in the memory.

Tip   QRAM consists of three components such as the memory array, input register, and output register. A memory array can be classical or quantum based. The registers consist of quantum bits.

Let's look at quantum sparse support vector machines in detail.

Quantum Sparse Support Vector Machines

There are various techniques such as least squares for simulating quantum sparse support vector machines. The Harrow–Hassidim–Lloyd (HHL) algorithm is used to find a solution for a linear equation–based system to forecast using trained quantum bits. This is a hybrid algorithm of the classical and quantum computing techniques.

Regularization is part of the sparse support vector machine technique. It is related to the penalty imposed on parameter weights in the objective function. A high-dimensional space consisting of features can be used to get accurate predictions. The L2 Norm penalty is used as a regularization algorithm, and L1 Norm is better than L2 Norm. The objective function in the support vector machine uses Hinge Loss minimization with L1 Norm. See Figure 9-7.

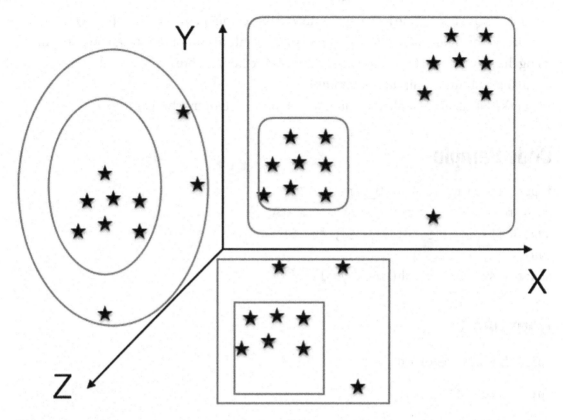

Figure 9-7. *Quantum support vector machine*

Linear programming is a form of semidefinite programming (SDP), which uses sparse support vector machine formulation. A quantum sparse support vector machine uses a quantum SDP solver. The algorithm steps for solving SDP are shown here:

1. Parameters are created in a dictionary.

2. The data set is created as an input object.

3. The algorithm executes based on the input data set and parameters.

4. The result with success ratio and details are printed.

5. The predicted labels are printed in the output.

Tip   A positive semidefinite matrix is a Hermitian matrix, which has positive eigenvalues.

ReNomQ is a machine learning framework based on quantum principles. Quantum circuits can be modeled using this framework. Quantum computation can be performed using the software package. A quantum support vector machine is modeled using an adiabatic quantum computation model.

Let's look at an example for a quantum support vector machine.

Code Sample

```
import renom_q.ml.quantum_svm
quantum_svm = renom_q.ml.quantum_svm.QSVM()
xval = [[3, 4], [5, 6], [7, 9], [8, 6]]
yval = [1, -1, 1, -1]
quantum_svm.plot_graph(xval, yval)
```

Command

```
pip3 install renorm_q

python3 quantum_svm.py
```

Output

Figure 9-8 shows the output.

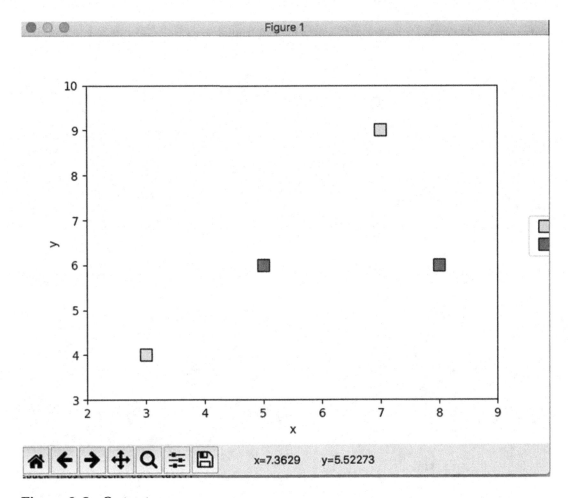

Figure 9-8. *Output*

Summary

In this chapter, we looked at techniques related to quantum classifier algorithms such as quantum classifiers, variational quantum classification, quantum support vector machines, and quantum sparse vector machines.

CHAPTER 10

Quantum Data Processing

Introduction

"Quantum attention functions are the keys to quantum machine learning."

—Amit Ray

This chapter covers the topic of quantum data processing (see Figure 10-1). You will see how quantum data processing algorithms are implemented in day-to-day life. Code examples are presented for algorithms such as k-means, k-medians, and quantum clustering.

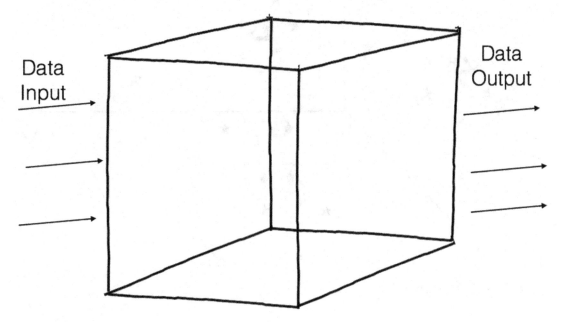

Figure 10-1. Data processing

© Bhagvan Kommadi 2020
B. Kommadi, *Quantum Computing Solutions*, https://doi.org/10.1007/978-1-4842-6516-1_10

Initial Setup

You need to set up Python 3.5 to run the code samples in this chapter. You can download it from https://www.python.org/downloads/release/python-350/.

Classical K-Means

The classical k-means technique partitions the input data into k clusters. The k clusters need to be different, and every data point is part of a single cluster only. A cluster is selected based on the minimum sum of the squared distance between the centroid and the datapoint. See Figure 10-2.

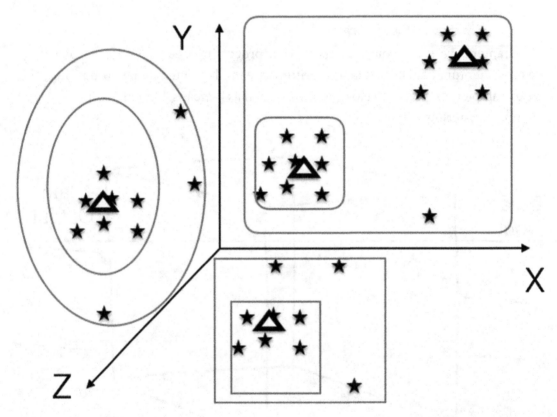

Figure 10-2. *K-means*

The pseudo-algorithm's steps are shown here:

1. Select k centroids after initialization in a random fashion.

2. A centroid is selected for every data point based on the minimum distance between the data point and the centroid.

3. The centroids for every cluster are recalculated based on the average of the data points in a cluster.

4. The previous two steps are repeated.

The data set needs to be chosen in such a way that the average is 0 and the standard deviation is 1. This technique works in different clusters of centroids selected initially, and the calculation is based on the k-means algorithm. The execution iterations are compared to select the best possible solution. This technique is used when the number of cluster centroids is associated with different types in the data set. This technique's performance is better when the variables are many. The challenge in this method is to forecast the k value.

Tip   In k-means, *k* stands for the number of clusters in the input data.

Let's look at the code implementation of the k-means algorithm.

Code Sample

```
import matplotlib.pyplot as plot
import seaborn as seasns; seasns.set()
import numpy as nump
from sklearn.cluster import KMeans as skKMeans

from sklearn.datasets.samples_generator import make_blobs
X, y_true = make_blobs(n_samples=250, centers=4,
                       cluster_std=0.60, random_state=0)
plot.scatter(X[:, 0], X[:, 1], s=50);

plot.show()
```

```
kmeans_res = skKMeans(n_clusters=4)
kmeans_res.fit(X)
y_val_kmeans = kmeans_res.predict(X)

plot.scatter(X[:, 0], X[:, 1], c=y_val_kmeans, s=50, cmap='viridis')

centers = kmeans_res.cluster_centers_

plot.scatter(centers[:, 0], centers[:, 1], c='grey', s=200, alpha=0.5);

plot.show()
```

Command

```
pip3 install seaborn
pip3 install numpy
python3 k-means.py
```

Output

The sample data is plotted and presented in the graph shown in Figure 10-3.

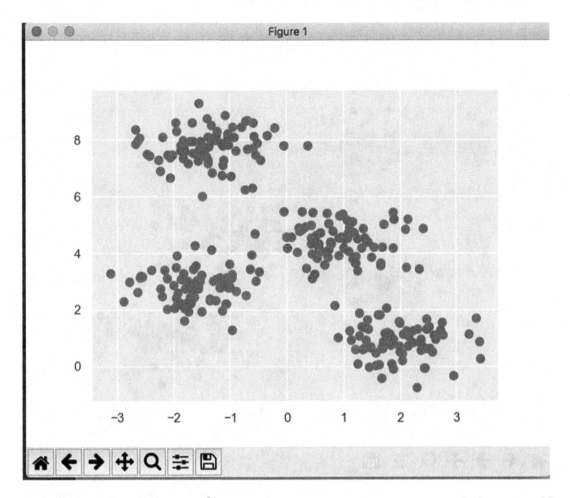

Figure 10-3. *K-means sample*

The k-means algorithm is applied, and the plot is presented in the graph in Figure 10-4.

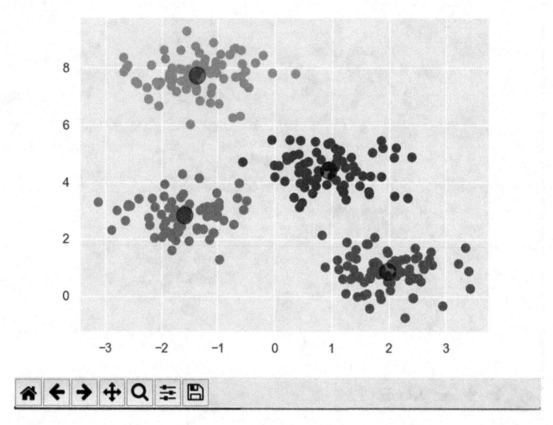

Figure 10-4. *K-means results*

Let's now look at the quantum k-means data processing technique.

Quantum K-Means

The quantum k-means technique (see Figure 10-5) is applied using a quantum circuit that can have training vectors. The training vectors are the initial centroids, and the centroids are computed after every step. The circuit can be changed by using a rotated training vector. The number of training vectors is increased based on the number of clusters. The pseudo-algorithm steps are shown here:

1. Input the k cluster groups, and input the data points.

2. Initialize the k centroids randomly.

3. Iterate all the input data points to identify the cluster group to them.

4. Compute the quantum interference circuit.

5. Filter the interference probabilities.

6. Add the high probability centroids into the dictionary.

7. The previous three methods are computed for the set of data points.

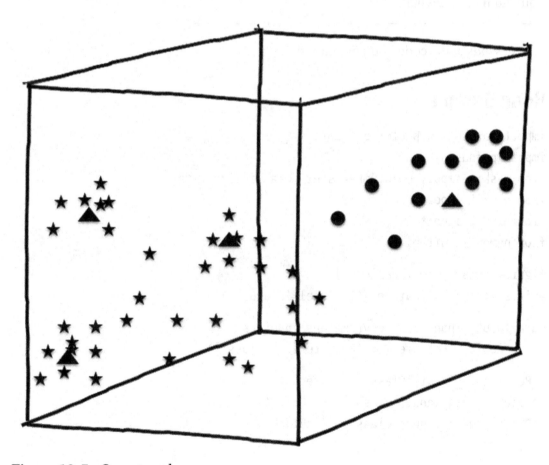

Figure 10-5. *Quantum k-means*

For k clusters, the N number of quantum machines can be modeled for the quantum k-means technique. The quantum state consists of the group of centroids that are the result of the algorithm. Parallelization can be done by breaking down the input data into different sets for each quantum machine to execute the algorithm. The quantum state is aggregated and used for iterations.

The Iris data set that was discussed in the previous chapter can be clustered in three groups (setosa, versicolor, virginica) based on the petal's and sepal's width and height. The technique tries to identify the cluster groups with lesser variation within the cluster. The other application is to classify people based on their income and spending into k clusters.

Tip   Trace distance and fidelity are the quantum distance measures used in quantum computation.

Let's look at the code implementation.

Code Sample

```
import matplotlib.pyplot as plot
import pandas as pand
from qiskit import QuantumRegister, ClassicalRegister
from qiskit import QuantumCircuit
from qiskit import Aer, execute
from numpy import pi

figure, axis = plot.subplots()
axis.set(xlabel='Feature 1', ylabel='Feature 2')

data_input = pand.read_csv('kmeans_input.csv',
    usecols=['Feature 1', 'Feature 2', 'Class'])

isRed = data_input['Class'] == 'Red'
isGreen = data_input['Class'] == 'Green'
isBlack = data_input['Class'] == 'Black'

# Filter data
redData = data_input[isRed].drop(['Class'], axis=1)
```

```python
greenData = data_input[isGreen].drop(['Class'], axis=1)
blackData = data_input[isBlack].drop(['Class'], axis=1)

y_p = 0.141
x_p = -0.161

xgc = sum(redData['Feature 1']) / len(redData['Feature 1'])
xbc = sum(greenData['Feature 1']) / len(greenData['Feature 1'])
xkc = sum(blackData['Feature 1']) / len(blackData['Feature 1'])

# Finding the y-coords of the centroids
ygc = sum(redData['Feature 2']) / len(redData['Feature 2'])
ybc = sum(greenData['Feature 2']) / len(greenData['Feature 2'])
ykc = sum(blackData['Feature 2']) / len(blackData['Feature 2'])

# Plotting the centroids
plot.plot(xgc, ygc, 'rx')
plot.plot(xbc, ybc, 'gx')
plot.plot(xkc, ykc, 'kx')

plot.plot(x_p, y_p, 'bo')

# Setting the axis ranges
plot.axis([-1, 1, -1, 1])

plot.show()

# Calculating theta and phi values
phi_list = [((x + 1) * pi / 2) for x in [x_p, xgc, xbc, xkc]]
theta_list = [((x + 1) * pi / 2) for x in [y_p, ygc, ybc, ykc]]

quantumregister = QuantumRegister(3, 'quantumregister')

classicregister = ClassicalRegister(1, 'classicregister')

quantum_circuit = QuantumCircuit(quantumregister, classicregister, name='qc')

backend = Aer.get_backend('qasm_simulator')

quantum_results_list = []

for i in range(1, 4):
    quantum_circuit.h(quantumregister[2])
```

```
    quantum_circuit.u3(theta_list[0], phi_list[0], 0, quantumregister[0])
    quantum_circuit.u3(theta_list[i], phi_list[i], 0, quantumregister[1])

    quantum_circuit.cswap(quantumregister[2], quantumregister[0],
    quantumregister[1])
    quantum_circuit.h(quantumregister[2])

    quantum_circuit.measure(quantumregister[2], classicregister[0])

    quantum_circuit.reset(quantumregister)

    job = execute(quantum_circuit, backend=backend, shots=1024)
    result = job.result().get_counts(quantum_circuit)
    quantum_results_list.append(result['1'])

print(quantum_results_list)

class_list = ['Red', 'Green', 'Black']

quantum_p_class = class_list[quantum_results_list.index(min(quantum_
results_list))]

distances_list = [((x_p - i[0])**2 + (y_p - i[1])**2)**0.5 for i in [(xgc,
ygc), (xbc, ybc), (xkc, ykc)]]
classical_p_class = class_list[distances_list.index(min(distances_list))]

print("""using quantumdistance algorithm,
 the new data point belongs to the""", quantum_p_class,
 'class.\n')
print('Euclidean distances are listed: ', distances_list, '\n')
print("""based on euclidean distance calculations,
 the new data point is related to the""", classical_p_class,
 'class.')
```

Command

```
pip3 install qiskit

pip3 install qiskit-ibmq-provider

pip3 install pandas
```

```
pip3 install matlab
```

```
python3 quantum_k_means.py
```

Output

The input points are read from the Excel file that has the colors red, green, and black. A new point is added to use the quantum k-means algorithm to find which class color it belongs to. Figure 10-6 shows the output.

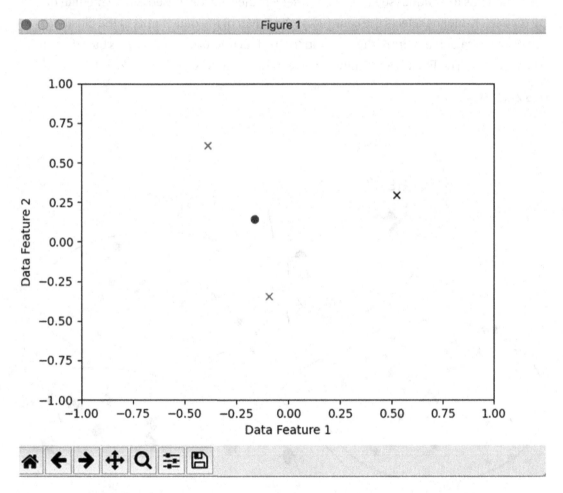

Figure 10-6. *Output*

Using the quantum distance algorithm, the new data point [72, 60, 125] is related to the Green class. The distances measured are based on Euclidean distances.

$$[0.520285324797846, 0.4905204028376393, 0.7014755294377704]$$

Now, let's look at the classic k-medians.

Classic K-Medians

The k-medians technique (see Figure 10-7) selects the centroid based on the median of the data set. K-means selects the centroid as the mean of the data points of the cluster. The distance calculation from the centroid to the data points in the cluster is based on the taxicab metric. The Euclidean distance between two points (x1,y1) and (x2, y2) is as follows:

$$\sqrt{(x2-x1)^2+(y2-y1)^2}$$

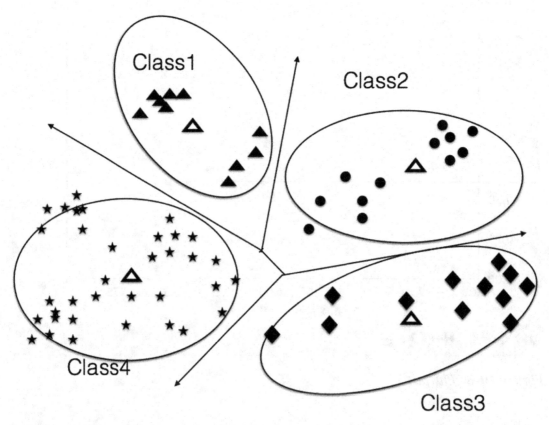

Figure 10-7. *K-medians*

The taxicab metric is as shown here:

```
|x2-x1| + |y2-y1|
```

Tip   The difference between k-medians and k-means is in the choice of the center of the cluster. K-means selects the centroid, and k-medians chooses the average of cluster points.

Let's look at the code implementation of the k-medians algorithm.

Code Sample

```python
import numpy as nump
from k_medians_utils import *

PREFIX = ''

FILE = PREFIX + 'iris.data.txt'

[xval, yval] = read(FILE)

[n, d] = nump.shape(xval)
k = len(nump.unique(yval))

mi = nump.min(xval, axis=0)
ma = nump.max(xval, axis=0)
di = ma - mi
stop = 0

c = nump.zeros(n)
median = nump.random.rand(k, d) * nump.ones((k, d))
median = median * di
median = median + mi
med_t = nump.copy(median)

imax = 100
for i in range(imax):
    med_t = nump.copy(median)
```

```python
    for j in range(n):
        dist = nump.sqrt(nump.sum(nump.power(xval[j,:] - median, 2), axis=1))
        idx = nump.argmin(dist)
        val = nump.min(dist)
        c[j] = idx

    for j in range(k):
        a = nump.arange(n)
        idx = a[c == j]
        l = len(idx)
        if l:
            median[j,:] = nump.median(xval[idx,:], axis=0)
        else:
            median[j,:] = median[j,:] + (nump.random.rand(d) * di)

    stop = nump.sum(nump.sum(nump.power(median - med_t,2), axis=0))
    if(stop <= 0) or (i >= imax):
        break

accur = getAccuracy(c, yval, k)
silhou = getSilhouette(xval, c, median)
print(accur, silhou)
```

k medians utils.py

```python
import numpy as nump
import os

VERBOSE = False

def read(FILE):
    if os.path.isfile(FILE):
        file = open(FILE, 'r')
        lines = tuple(file)
        file.close()
        data = []
        for line in lines:
            data.append(line.rstrip().split(","))
```

```python
        if VERBOSE:
            print(data[-1])
    else:
        print(FILE, 'does not exist')
        exit(0)

    data = nump.array(data)
    x = data[:,0:-1]
    x = x.astype(nump.float)

    y = nump.zeros(len(data))
    uniq = nump.unique(data[:,-1])
    for i in range(0,len(uniq)):
        idx = (data[:,-1] == uniq[i])
        if any(idx):
            y[idx] = i

    return(x, y)

def getAccuracy(c, y, k):
    if VERBOSE:
        print(c)
        print(y)

    n = len(y)
    kk = nump.zeros(k)
    o = 0
    e = 0
    idxx = []
    for i in range(k):
        a = nump.arange(n)
        idxa = a[y == i]
        for j in range(len(idxa)):
            kk[int(c[j+o])] = kk[int(c[j+o])] + 1
        if idxx:
            for l in idxx:
                kk[l] = 0
        o = o + len(idxa)
```

```
        idx = nump.argmax(kk)
        idxx.append(idx)
        val = kk[idx]
        e = e + (val/len(y[y == i]))
        kk = nump.zeros(k)
    e = e/k
    return(e)

def accuracy_(c, y, k1, k2):
    if VERBOSE:
        print(c)
        print(y)

    n = len(y)
    kk = nump.zeros(k2)
    o = 0
    e = 0
    idxx = []
    for i in range(k1):
        a = nump.arange(n)
        idxa = a[y == i]
        for j in range(len(idxa)):
            kk[int(c[j+o])] = kk[int(c[j+o])] + 1
        if idxx:
            for l in idxx:
                kk[l] = 0
        o = o + len(idxa)
        idx = nump.argmax(kk)
        idxx.append(idx)
        val = kk[idx]
        e = e + (val/len(y[y == i]))
        kk = nump.zeros(k2)
    e = e/k1
    return(e)
```

```python
def getSilhouette(x, c, me):
    if VERBOSE:
        print(c)
        print(y)

    n = len(c)
    s = nump.zeros((n,3))
    for i in range(n):
        dist = nump.sqrt(nump.sum(nump.power(x[i,:] - me,2), axis=1))
        dd = nump.argsort(dist)
        aa = nump.arange(n)
        for j in range(2):
            aa = nump.arange(n)
            idx = aa[c == dd[j]]
            l = len(idx)
            if l:
                for o in idx:
                    s[o,j] = s[o,j] + nump.sqrt(nump.sum(nump.power(x[i,:] -
                    x[o,:] ,2)))
                s[o,j] = s[o,j]/l
    s = nump.mean((s[:,0] - s[:,1])/nump.amax(s[:,0:2], axis=1))
    return(s)

def actionselection(action, probability, numactions, numdims):
    for i in range(numdims):
        a = nump.random.choice(nump.arange(0, numactions),
        p = probability[:,i])
        mask = nump.zeros(numactions,dtype=bool)
        #print(i, a, mask)
        mask[a] = True
        action[mask,i] = 1
        action[~mask,i] = 0
    return(action)
```

```
def probabilityupdate(action, probability, numactions, numdims, signal,
alpha, beta):
    if(numactions > 1):
        for i in range(numdims):
            a = nump.where(action[:,i] == 1)
            mask = nump.zeros(numactions,dtype=bool)
            mask[a] = True

            if not signal[i]:
                probability[mask,i] = probability[mask,i] + alpha *
                (1 - probability[mask,i])
                probability[~mask,i] = (1 - alpha) * probability[~mask,i]
            else:
                probability[mask,i] = (1 - beta) * probability[mask,i]
                probability[~mask,i] = (beta/(numactions-1)) + (1-beta) *
                probability[~mask,i]
    return(probability)
```

Command

```
pip3 install nump
```

```
python3 k-medians.py
```

Output

```
0.9 -0.2609494581992614
```

The iris data set that was discussed in the previous chapter can be clustered into three different groups (setosa, versicolor, virginica) based on the petal and sepal's width and height. The k-medians clustering algorithm has an accuracy of 90 percent in clustering the data points to the setosa, versicolor, and virginica groups.

Let's look at quantum k-medians in detail.

Quantum K-Medians

The quantum k-medians technique is the computation of the cluster centroids in a cluster group by using the median of the data points. A median is computed based on the minimal distance from all the data points. The technique is related to minimizing the sum of the distance between a point and the median of the data set. The minimization of the distance is computed based on a quantum minimization algorithm using Grover's search. The Euclidean distance between the points is used as the distance algorithm.

The steps for the pseudo-algorithm are shown here:

1. The input is the k cluster groups and input data points.

2. Initialize the k centroids randomly based on the median of the data points.

3. All the input data points are iterated to identify the cluster group to them.

4. The quantum interference circuit is computed.

5. The interference probabilities are filtered.

6. The high probability centroids are added into the dictionary.

7. The previous three methods are computed for the set of data points.

Tip   Grover's search algorithm uses a unitary oracle function to create a superposition state based on a function f(x).

Now let's look at the classical clustering algorithm.

Classical Clustering

Classical clustering techniques are used to separate the input data into multiple groups based on a set of features. A group will have data points that have commonality. Multiple groups are created to handle different data points. This is based on unsupervised machine learning techniques. Clustering identifies patterns and features in a group of unlabeled input data.

These are the different types of classical clustering techniques:

- Partition based

- Hierarchical

- Density based

- Grid based

- Model based

The classical clustering technique is used in image processing, computational life sciences, healthcare, and economics applications. See Figure 10-8.

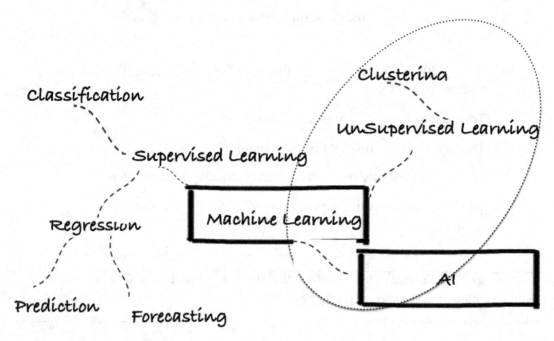

Figure 10-8. *Clustering*

Let's start looking at quantum clustering techniques.

Quantum Clustering

The quantum clustering algorithm (see Figure 10-9) is based on the gradient descent technique. It is used to find the quantum potential at a constant learning rate and computing the cluster center. Using quantum mechanics–based principles, quantum clustering techniques find groups with complex shapes. The technique identifies the group of any shape by computing the group center.

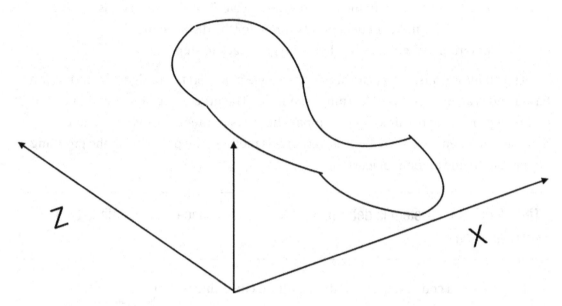

Figure 10-9. *Quantum clustering*

Quantum clustering is based on the inversion problem in quantum mechanics. The quantum clustering technique helps to get a particle distribution that is estimated based on the potential function. The algorithm finds the group center. For each group in the center, it assigns the center. The wave function is identified from the Schrodinger equation solution.

Let's look at the steps for the quantum clustering algorithm:

1. The weights for each data feature are identified, and the parameters are selected based on the input data.

2. The number of groups/measurement scale is set to zero.

3. The weighted measures are calculated based on the quantum clustering distance method.

4. The parameter and the potential energy based on the data are estimated.

5. The group number is increased by 1.

6. The minimum potential energy is identified at the group's center.

7. All the data points are grouped by the distance metric, and it needs to be less than the measurement scale. The algorithm ends when the number of data points is zero for the distance metric criterion. Otherwise, the algorithm goes back to step 5.

The quantum clustering algorithm has a single parameter. The algorithm is partition based and is an unsupervised learning technique. The processing time is higher because of the preprocessing required for grouping. The measurement scale is static, and the method is not dependent on the features. In this method, the precision of the grouping is dependent on the amount of learning.

Tip   A quantum system is deterministic in time and space and described by a wave function.

Let's look at a code example for the quantum clustering technique.

Code Sample

```
import numpy as nump
import quantum_cluster

types = ['f8', 'f8', 'f8','f8','U50']

data_res = nump.loadtxt('iris_input.csv', delimiter=',')

print(data_res)
data_res = data_res[:,:4]

sigma_val=0.55
repetitionsVal=100
stepSizeVal=0.1
clusteringTypeVal='v'
isRecalculate=False
isReturnHistory=True
isStopCondition=True
voxelSizeVal = None
```

```
xval,xHistoryVal = quantum_cluster.getGradientDescent(data_res,sigma=sigma_
val,repetitions=repetitionsVal,stepSize=stepSizeVal,clusteringType=clusteri
ngTypeVal,recalculate=isRecalculate,returnHistory=isReturnHistory,stopCondi
tion=isStopCondition,voxelSize=voxelSizeVal)

clusters_res = quantum_cluster.PerformFinalClusteringAlgo(xval,stepSizeVal)

quantum_cluster.displayClusteringValues(xHistoryVal,clusters_res)
```

quantum_cluster.py

```
import numpy as nump
from scipy.spatial import distance as spatDistance
from sklearn.decomposition import PCA as decomPCA
import matplotlib.pyplot as plot
from mpl_toolkits.mplot3d import Axes3D as mplAxes3D

def getVGradient(data:nump.ndarray,sigma,x:nump.ndarray=None,coeffs:nump.
ndarray=None):

    if x is None:
        x = data.copy()

    if coeffs is None:
        coeffs = nump.ones((data.shape[0],))

    twoSigmaSquared = 2*sigma**2

    data = data[nump.newaxis,:,:]
    x = x[:,nump.newaxis,:]
    differences = x-data
    squaredDifferences = nump.sum(nump.square(differences),axis=2)
    gaussian = nump.exp(-(1/twoSigmaSquared)*squaredDifferences)
    laplacian = nump.sum(coeffs*gaussian*squaredDifferences,axis=1)
    parzen = nump.sum(coeffs*gaussian,axis=1)
    v = 1 + (1/twoSigmaSquared)*laplacian/parzen

    dv = -1*(1/parzen[:,nump.newaxis])*nump.sum(differences*((coeffs*ga
ussian)[:,:,nump.newaxis])*(twoSigmaSquared*(v[:,nump.newaxis,nump.
newaxis])-(squaredDifferences[:,:,nump.newaxis])),axis=1)
```

213

```
    v = v-1

    return v, dv

def getSGradient(data:nump.ndarray,sigma,x:nump.ndarray=None,coeffs:nump.
ndarray=None):

    if x is None:
        x = data.copy()

    if coeffs is None:
        coeffs = nump.ones((data.shape[0],))

    twoSigmaSquared = 2 * sigma ** 2

    data = data[nump.newaxis, :, :]
    x = x[:, nump.newaxis, :]
    differences = x - data
    squaredDifferences = nump.sum(nump.square(differences), axis=2)
    gaussian = nump.exp(-(1 / twoSigmaSquared) * squaredDifferences)
    laplacian = nump.sum(coeffs*gaussian * squaredDifferences, axis=1)
    parzen = nump.sum(coeffs*gaussian, axis=1)
    v = (1 / twoSigmaSquared) * laplacian / parzen
    s = v + nump.log(nump.abs(parzen))

    ds = (1 / parzen[:, nump.newaxis]) * nump.sum(differences *
    ((coeffs*gaussian)[:, :, nump.newaxis]) * (
    twoSigmaSquared * (v[:, nump.newaxis, nump.newaxis]) -
    (squaredDifferences[:, :, nump.newaxis])), axis=1)

    return s, ds

def getPGradient(data:nump.ndarray,sigma,x:nump.ndarray=None,coeffs:nump.
ndarray=None):

    if x is None:
        x = data.copy()

    if coeffs is None:
        coeffs = nump.ones((data.shape[0],))

    twoSigmaSquared = 2 * sigma ** 2
```

```python
    data = data[nump.newaxis, :, :]
    x = x[:, nump.newaxis, :]
    differences = x - data
    squaredDifferences = nump.sum(nump.square(differences), axis=2)
    gaussian = nump.exp(-(1 / twoSigmaSquared) * squaredDifferences)
    p = nump.sum(coeffs*gaussian,axis=1)

    dp = -1*nump.sum(differences * ((coeffs*gaussian)[:, :, nump.newaxis])
    * twoSigmaSquared,axis=1)

    return p, dp

def getApproximateParzenValues(data:nump.ndarray,sigma,voxelSize):

    newData = getUniqueRows(nump.floor(data/voxelSize)*voxelSize+
    voxelSize/2)[0]

    nMat = nump.exp(-1*spatDistance.squareform(nump.square(spatDistance.
    pdist(newData)))/(4*sigma**2))
    mMat = nump.exp(-1 * nump.square(spatDistance.cdist(newData,data)) /
    (4 * sigma ** 2))
    cMat = nump.linalg.solve(nMat,mMat)
    coeffs = nump.sum(cMat,axis=1)
    coeffs = data.shape[0]*coeffs/sum(coeffs)

    return newData,coeffs

def getUniqueRows(x):
    y = nump.ascontiguousarray(x).view(nump.dtype((nump.void, x.dtype.
    itemsize * x.shape[1])))
    _, inds,indsInverse,counts = nump.unique(y, return_index=True,return_
    inverse=True,return_counts=True)

    xUnique = x[inds]
    return xUnique,inds,indsInverse,counts

def getGradientDescent(data,sigma,repetitions=1,stepSize=None,clusteringTy
pe='v',recalculate=False,returnHistory=False,stopCondition=True,voxelSize
=None):
```

```python
n = data.shape[0]

useApproximation = (voxelSize is not None)

if stepSize is None:
    stepSize = sigma/10

if clusteringType == 'v':
    gradientFunction = getVGradient
elif clusteringType == 's':
    gradientFunction = getSGradient
else:
    gradientFunction = getPGradient

if useApproximation:
    newData, coeffs = getApproximateParzenValues(data, sigma,
    voxelSize)
else:
    coeffs = None

if recalculate:
    if useApproximation:
        x = nump.vstack((data,newData))
        data = x[data.shape[0]:]
    else:
        x = data
else:
    if useApproximation:
        x = data
        data = newData
    else:
        x = data.copy()

if returnHistory:
    xHistory = nump.zeros((n,x.shape[1],repetitions+1))
    xHistory[:,:,0] = x[:n,:].copy()

if stopCondition:
    prevX = x[:n].copy()
```

```
    for i in range(repetitions):
        if ((i>0) and (i%10==0)):
            if stopCondition:
                if nump.all(nump.linalg.norm(x[:n]-prevX,axis=1) < nump.
                sqrt(3*stepSize**2)):
                    i = i-1
                    break
                prevX = x[:n].copy()

        f,df = gradientFunction(data,sigma,x,coeffs)
        df = df/nump.linalg.norm(df,axis=1)[:,nump.newaxis]
        x[:] = x + stepSize*df

        if returnHistory:
            xHistory[:, :, i+1] = x[:n].copy()

    x = x[:n]

    if returnHistory:
        xHistory = xHistory[:,:,:(i+2)]
        return x,xHistory
    else:
        return x

def PerformFinalClusteringAlgo(data,stepSize):
    clusters = nump.zeros((data.shape[0]))
    i = nump.array([0])
    c = 0
    spatDistances = spatDistance.squareform(spatDistance.pdist(data))
    while i.shape[0]>0:
        i = i[0]
        inds = nump.argwhere(clusters==0)
        clusters[inds[spatDistances[i,inds] <= 3*stepSize]] = c
        c += 1
        i = nump.argwhere(clusters==0)
    return clusters

def displayClusteringValues(xHistory,clusters=None):
```

```python
plot.ion()
plot.figure(figsize=(20, 12))
if clusters is None:
    clusters = nump.zeros((xHistory.shape[0],))
if xHistory.shape[1] == 1:

    sc = plot.scatter(xHistory[:,:,0],xHistory[:,:,0]*0,c=clusters
    ,s=10)
    plot.xlim((nump.min(xHistory),nump.max(xHistory)))
    plot.ylim((-1,1))
    for i in range(xHistory.shape[2]):
        sc.set_offsets(xHistory[:, :, i])
        plot.title('step #' + str(i) + '/' + str(xHistory.shape[2]-1))
        plot.pause(0.05)
elif xHistory.shape[1] == 2:

    sc = plot.scatter(xHistory[:, 0, 0], xHistory[:, 1, 0] ,
    c=clusters, s=20)
    plot.xlim((nump.min(xHistory[:,0,:]), nump.max(xHistory[:,0,:])))
    plot.ylim((nump.min(xHistory[:, 1, :]), nump.max(xHistory
    [:, 1, :])))
    for i in range(xHistory.shape[2]):
        sc.set_offsets(xHistory[:, :, i])
        plot.title('step #' + str(i) + '/' + str(xHistory.shape[2]-1))
        plot.pause(0.2)
else:
    if xHistory.shape[1] > 3:
        pca = decomPCA(3)
        pca.fit(xHistory[:,:,0])
        newXHistory = nump.zeros((xHistory.shape[0],3,
        xHistory.shape[2]))
        for i in range(xHistory.shape[2]):
            newXHistory[:,:,i] = pca.transform(xHistory[:,:,i])
        xHistory = newXHistory

    ax = plot.axes(projection='3d')
```

```
sc = ax.scatter(xHistory[:, 0, 0], xHistory[:, 1, 0],xHistory
[:, 2, 0], c=clusters, s=20)
ax.set_xlim((nump.min(xHistory[:, 0, :]), nump.max(xHistory
[:, 0, :])))
ax.set_ylim((nump.min(xHistory[:, 1, :]), nump.max(xHistory
[:, 1, :])))
ax.set_zlim((nump.min(xHistory[:, 2, :]), nump.max(xHistory
[:, 2, :])))
for i in range(xHistory.shape[2]):
    sc._offsets3d =  (nump.ravel(xHistory[:, 0, i]),nump.
    ravel(xHistory[:, 1, i]),nump.ravel(xHistory[:, 2, i]))
    plot.gcf().suptitle('step #' + str(i) + '/' + str(xHistory.
    shape[2]-1))
    plot.pause(0.01)

plot.ioff()
plot.close()
```

Command

```
pip3 install nump
python3 QCAlgo.py
```

Output

Using the quantum clustering algorithm, you can create clusters. Figure 10-10 shows the output after the fourth step.

step #4/50

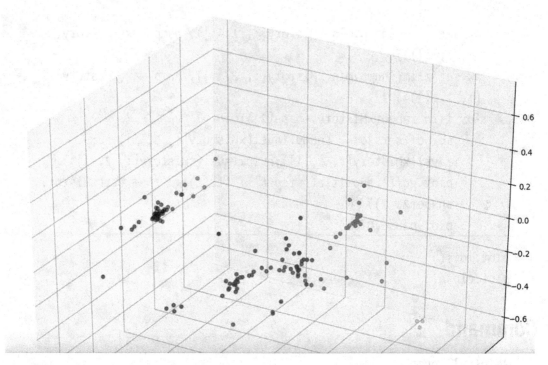

Figure 10-10. *Quantum clustering output after four steps*

Figure 10-11 shows the output after 29 steps, and Figure 10-12 shows the output after 46 steps.

step #29/50

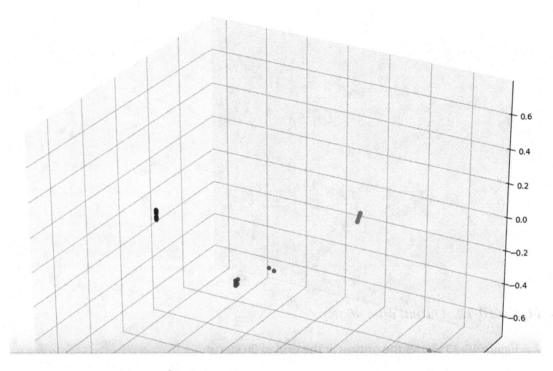

Figure 10-11. *Output after 29 steps*

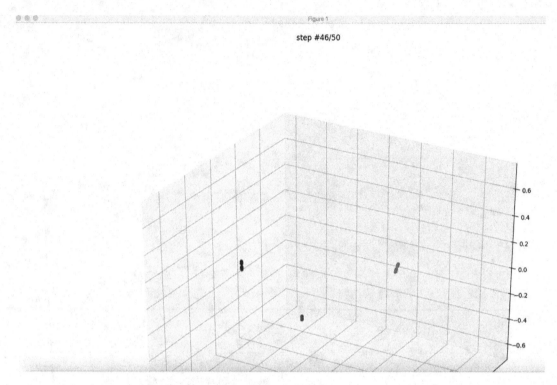

Figure 10-12. *Output after 46 steps*

Figure 10-13 shows the output at the end of 50 steps.

step #50/50

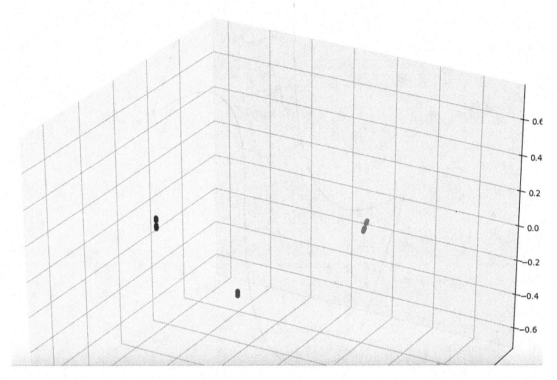

Figure 10-13. *Output after 50 steps*

Quantum Manifold Embedding

Quantum manifold embedding (see Figure 10-14) has sections that are Euclidean spaces. Euclidean spaces represent the quantum state in a quantum system. The Euclidean space has m vectors related to m dimensions. A classical manifold is an m-dimensional shape embedded in a higher-dimensional Euclidean system. Isomap is an example of a manifold embedding algorithm. Isomap helps to find a global optimum using the manifold underneath.

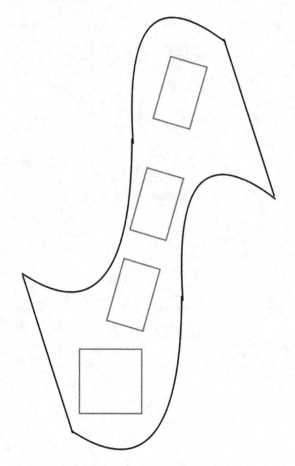

Figure 10-14. *Manifold embedding*

Summary

In this chapter, we looked at the techniques related to quantum data processing algorithms, such as quantum k-means, quantum k-medians, quantum clustering, and quantum manifold embedding.

CHAPTER 11

Quantum AI Algorithms

Introduction

"If we take quantum theory seriously as a picture of what's really going on, each measurement does more than disturb: it profoundly reshapes the very fabric of reality."

—Nick Herbert

This chapter covers the topic of quantum AI algorithms. You can see how quantum AI algorithms are implemented in day-to-day life. Code examples are presented for specific algorithms such as quantum probability, quantum walks, quantum search, quantum deep learning, and quantum parallelism.

Initial Setup

You need to set up Python 3.5 to run the code samples in this chapter. You can download it from `https://www.python.org/downloads/release/python-350/`.

Quantum Probability

A quantum computation framework has probability-based events that are modeled as Hilbert multidimensional subspaces. The system in context is referred to as a *state vector* in Hilbert space. Event probabilities are computed by using vector projection on subspaces and calculating the projection length squared.

Now, let's look in detail at Hilbert space. It is a mathematical vector space that has a dot product or an inner product of vectors. The space is related to the distance function, which is referred to as the *inner product space*.

© Bhagvan Kommadi 2020
B. Kommadi, *Quantum Computing Solutions*, https://doi.org/10.1007/978-1-4842-6516-1_11

Quantum probability (see Figure 11-1) is based on mathematical quantum theory. A quantum probability space represents a system that is not well known in physical space.

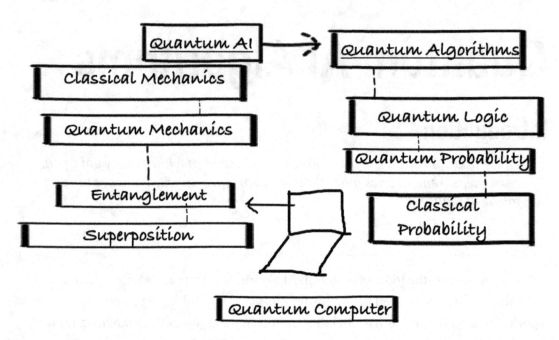

Figure 11-1. *Quantum probability*

A quantum probability space is a pair (B, ρ) where B is a Q-algebra and ρ : B → D is a function with the following properties:

(a) $\rho(aB + bC) = a\rho(B) + b\rho(C)$ (B, C ∈ A, a, b ∈ D),

(b) $\rho(B*) = \rho(B) * (B \in B)$,

(c) $\rho(B*B) \geq 0$ (B ∈ B)

A projection Q ∈ B is an observation on the system. A partition of the measurement that is an identity {Q1, ..., Qm} yields the measurements Q1, ..., Qm. The larger $\rho(Q)$ is, the more likely the measurement is Q.

The law of probability is based on $\rho(Q)$, where Q ∈ B. If $\rho(Q)=1$, the yield Q will be the measurement of the Q value.

Tip   Quantum probability is based on Von Neumann axioms and relaxed constraints, which are based on Kolmogorov probability.

Classic Walks

A classical random walkc (see Figure 11-2) is based on the behavior of a particle traversing on the line. The particle movement is based on the probability rules. The simplest walk of a particle is modeled as a time step with the unit distance moved left or right. The movement has probability q for left and 1-q for right. The probabilities are independent of the previous positions.

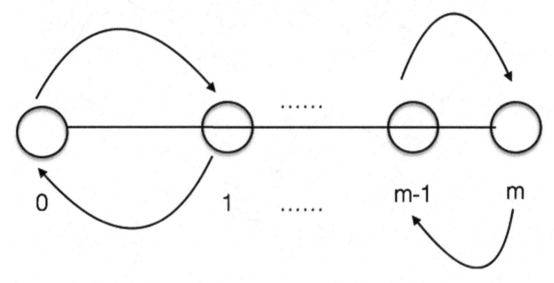

Figure 11-2. *Classical random walk*

Now let's look at the quantum walks technique.

Quantum Walks

Quantum walks (see Figure 11-3) are based on the traversal of the quantum particle, which is in a superposition state. The probabilities are associated with the left and right movements. The sum of the probabilities of the superposition states is not unitary. A quantum walk is based on a particle's spin. A random quantum walk is possible by rotating the particle with spin. Quantum walks are used for creating quantum algorithms based on the quantum computation model. There are two types of quantum walks: discrete-time and continuous-time quantum walks.

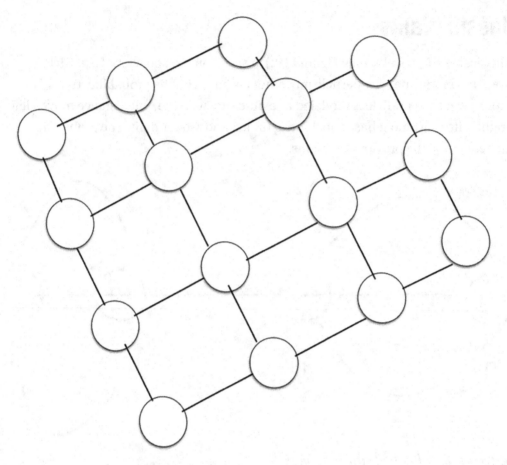

Figure 11-3. *Quantum random walk*

The coined walk model (see Figure 11-4) is based on the coin and identity operators of Hilbert space. The operator used in the coined walk model is D ⊗ M, where D is the coin operator that represents the particle's motion and M is the identity operator of the Hilbert space GP. The walk model is based on the coin toss. The computation model is based on the superposition of the quantum states modeling the coin state. The Hadamard operator is the frequently used coin. After applying the coin operator, a shift operator is applied. The superposition state is created by superposition of m=1 and m=-1. The probability distribution in the coined quantum walk is not normal distribution. The standard deviation of this quantum walk is not $\sqrt{s}$. It is proportional to s steps. Assume that the probability of the quantum particle to go to the right is 1. After s steps, the particle is found at m=s. This scenario is referred to as a *ballistic case*. The quantum particle is assumed to have unit velocity in motion. The speed of the quantum particle in a ballistic scenario is half the speed of the free particle.

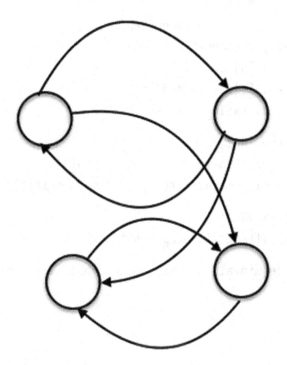

Figure 11-4. *Coined walk model*

Let's look at the code implementation of a coin toss quantum walk.

Code Sample

```
import math
import pylab as pythonLab

def GetProbabilities(posn):

    return [sum([abs(amp) ** 2 for amp in place]) for place in posn]

def ApplyNormalisation(posn):

    N = math.sqrt(sum(GetProbabilities(posn)))
    return [[amp / N for amp in place] for place in posn]

def GetTimeStep(posn):

    return ApplyNormalisation([[x[0] + x[1], x[0] - x[1]] for x in posn])

def GetShift(coin):
```

```
    newposn = [[0, 0] for i in range(len(coin))]
    for j in range(1, len(position) - 1):
        newposn[j + 1][0] += coin[j][0]
        newposn[j - 1][1] += coin[j][1]
    return ApplyNormalisation(newposn)

minval, maxval = -400, 401
position = [[0, 0] for i in range(minval, maxval)]
position[-minval] = [1 / math.sqrt(2), 1j / math.sqrt(2)]

for time in range(-minval):
    position = GetShift(GetTimeStep(position))

pythonLab.plot(range(minval, maxval), GetProbabilities(position))
pythonLab.show()
```

Command

```
pip3 install pylab

python3 quantum_walk.py
```

Output

Figure 11-5 shows the quantum walk output.

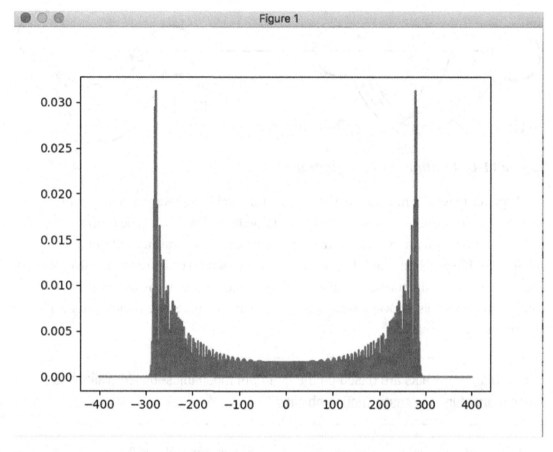

Figure 11-5. *Quantum walk output*

Let's now look at the continuous-time quantum walk.

A continuous-time quantum walk (see Figure 11-6) is based on the Markov process. A Markov process is modeled as a graph with M vertices that are connected to others with an edge. Assume γ is the jumping rate per unit time between the vertices. The constraint is that the walk is between the nodes on the edge. The walk is represented by stochastic generator matrix P. The sum of the overlap integrals squared across all vertices is 1.

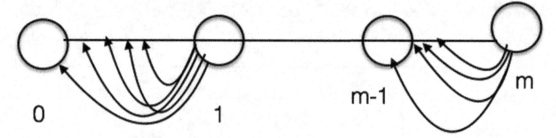

Figure 11-6. *Continuous-time quantum walk*

A quantum walk is modeled using a quantum mechanical wave function. The wave function is coherent superposition of all possible wave paths. The probability distribution is based on the quantum interference effects. Quantum walks help in solving real-life problems, and they are faster in polynomial order of time. The problems such as element distinctness, identifying triangle, and NAND tree evaluation are solved using quantum walks. Grover's search algorithm can be modeled as a quantum walk algorithm.

Tip   Quantum walks are based on the quantum mechanics–based walks that model quantum stochastic motion physics.

Let's look at the code implementation of a coin toss quantum walk.

Code Sample

```
from qutip import *
import numpy as nump
import matplotlib.pyplot as matplot
from math import *
import seaborn as seabrn

ket0 = basis(2,0).unit()
ket1 = basis(2,1).unit()
psip = (basis(2,0)+basis(2,1)*1j).unit()
psim = (basis(2,0)-basis(2,1)*1j).unit()

def GetCoin(coin_angle):
```

```
        C_hat = qutip.Qobj([[cos(radians(coin_angle)), sin(radians(coin_angle))],
                            [sin(radians(coin_angle)), -cos(radians(coin_angle))]])
        return C_hat

def GetShift(t):
        sites = 2*t+1
        shift_l = qutip.Qobj(nump.roll(nump.eye(sites), 1, axis=0))
        shift_r = qutip.Qobj(nump.roll(nump.eye(sites), -1, axis=0))
        S_hat = tensor(ket0*ket0.dag(),shift_l) + tensor(ket1*ket1.
        dag(),shift_r)
        return S_hat

def GetWalk(t,coin_angle):
        sites = 2*t+1
        C_hat = GetCoin(coin_angle)
        S_hat = GetShift(t)
        W_hat = S_hat*(tensor(C_hat,qeye(sites)))
        return W_hat

def ApplyDephasing(t,qstate,p_c):
        sites=2*t+1
        dstate = (1-p_c)*qstate+p_c*(tensor(sigmaz(),qeye(sites))*qstate*tensor
        (sigmaz(),qeye(sites)))
        return dstate

def ApplyDepolarization(t,qstate,p_c):
        sites=2*t+1
        dstate = (1-p_c)*qstate+p_c/3*(tensor(sigmax(),qeye(sites))*(qstate)*te
        nsor(sigmax(),qeye(sites))*
                            tensor(sigmay(),qeye(sites))*(qstate)*tensor(sig
                            may(),qeye(sites))*
                            tensor(sigmaz(),qeye(sites))*(qstate)*tensor(sig
                            maz(),qeye(sites)))
        return dstate

def GetQWalk_gen_markov(t,qubit_state,coin_angle,p_c):
        sites=2*t+1
        Position_state = basis(sites,t)
```

```
    Psi = ket2dm(tensor(qubit_state,Position_state))
    W_hat = GetWalk(t,coin_angle)
    for i in range(t):
        Psi = W_hat*ApplyDephasing(t,Psi,p_c)*W_hat.dag()
    return Psi

def GetMeasurement(t,Psi,z):
    sites=2*t+1
    prob=[]
    for i in range(0,sites,z):
        M_p = basis(sites,i)*basis(sites,i).dag()
        Measure = tensor(qeye(2),M_p)
        p = abs((Psi*Measure).tr())
        prob.append(p)
    return prob

def GetPlotParticlePos(P_p):
    lattice_positions = range(-len(P_p)+1,len(P_p)+1,2)
    matplot.plot(lattice_positions,P_p)
    matplot.xlim([-len(P_p)+1,len(P_p)+1])
    matplot.ylim([min(P_p),max(P_p)+0.01])
    matplot.ylabel(r'$Probablity Distribution$')
    matplot.xlabel(r'$Particle \ position$')
    matplot.show()

Psi_t = GetQWalk_gen_markov(150,psip,65,0.01)
P_p  = GetMeasurement(150,Psi_t,2)
GetPlotParticlePos(P_p)
```

Command

```
pip3 install git+https://github.com/qutip/qutip

python3 continuous_quantum_walk.py
```

Output

Figure 11-7 shows the output of a continuous-time quantum walk.

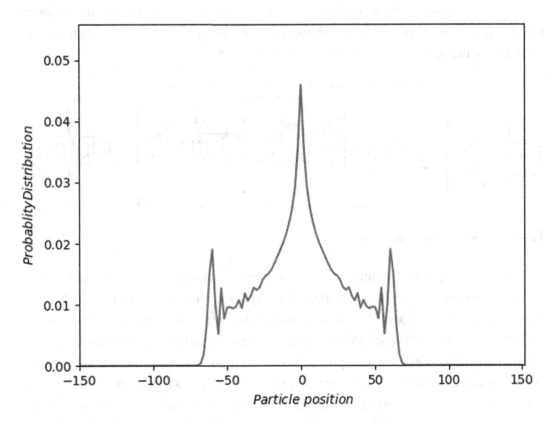

Figure 11-7. *Continuous quantum walk*

Now, let's look at the classic search algorithm.

Classic Search

In a classic search algorithm, an item is identified as having a particular attribute in a given set of M items. An item may or may not have the specified attribute. The search algorithm has complexity proportional to the count of queries to identify the particular item. These items are typically stored in a data store and split into different entities in a relational database. If the data store is NoSQL, they are stored in different collections. The items can be sorted or filtered based on an attribute or value, respectively.

Let's look at the quantum search algorithm.

Quantum Search

Quantum computers can be used for search algorithms execution. The superposition state helps to reduce complexity of search in quantum algorithms. The quantum search algorithms are proportional to /M queries. Quantum search is modeled using wave-like amplitudes and interference concepts. See Figure 11-8.

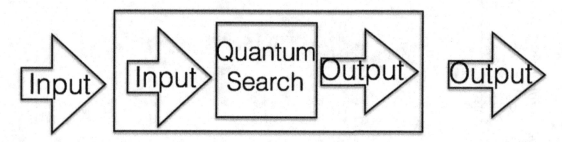

Figure 11-8. *Quantum search*

Let's now look at multi-element antenna arrays. In a multi-element antenna array, different phases can create radiation in a specific direction. The array analysis is performed using unitary transformations, and the amplitudes square results in the probabilities of the individual components. The transformations are based on the rotation and reflection of the components. This example is similar to quantum search. Quantum search identifies a sequence of quantum logic gates that finds a solution to a real-life problem. The algorithm is proportional to the amplitude of the particular time in context. Quantum search algorithms are popular and solve many real-life problems.

Tip   The quantum search algorithm is based on a quantum circuit that has Hadamard and CNOT gates.

Let's look at quantum deep learning in detail.

Quantum Deep Learning

Quantum algorithms are used to solve complex real-life problems. Deep learning is a popular technique used in quantum algorithms. The quantum computation framework, based on the Boltzmann machine, helps to provide a quantum deep learning framework. The framework helps in the process of learning and training the machine learning models.

Quantum machine learning is based on the quantum data and hybrid quantum classical models. The quantum data model is based on the quantum data source and uses the quantum computer–generated data. The model is based on the superposition and entanglement principles. A Hilbert space of 2^53 elements that represents joint probability distribution can be solved using a quantum computer. A Quantum Tensor Flow framework has the infrastructure concepts to create machine learning models that can use the quantum data models and algorithms. Quantum data models can be used in chemical simulation, quantum matter modeling, quantum control, quantum communication networks, and quantum metrology.

Hybrid quantum classical models are used for data generalization and solved using hybrid quantum classical techniques. A quantum neural network is one of the techniques that is based on the parameterized quantum computational model.

Tip   Quantum deep learning is used in the area of quantum pattern recognition for classification.

Quantum machine learning is applied in real-life scenarios. For instance, quantum chemistry problems are related to optimizing high-dimensional and difficult cost functions. The quantum computing framework is applied to solve computationally challenging problems with methods such as the variational quantum eigensolver.

A variational quantum eigensolver is about identifying the eigenvalues of certain operators. Eigen value identification is a complex and challenging problem. The quantum phase estimation method can be applied to compute the eigenvalue of a given eigenvector. You can optimize a variational circuit in lowering the squared energy expectation of a user-defined Hamiltonian. The Hamiltonian can be expressed as a sum of two Pauli operators.

Code Sample

```
import pennylane as quantumDeepL
from pennylane import numpy as npython
from pennylane.optimize import GradientDescentOptimizer as GDOpt
dev = quantumDeepL.device('default.qubit', wires=2)
```

```
def FindAnsatz():
    quantumDeepL.Rot(0.3, 1.8, 5.4, wires=1)
    quantumDeepL.RX(-0.5, wires=0)
    quantumDeepL.RY( 0.5, wires=1)
    quantumDeepL.CNOT(wires=[0, 1])
@quantumDeepL.qnode(dev)
def FindCircuitX():
    FindAnsatz()
    return quantumDeepL.expval(quantumDeepL.PauliX(1))
@quantumDeepL.qnode(dev)
def FindCircuitY():
    FindAnsatz()
    return quantumDeepL.expval(quantumDeepL.PauliY(1))
def getCost(var):
    expX = FindCircuitX()
    expY = FindCircuitY()
    return (var[0] * expX + var[1] * expY) ** 2
optimizer = GDOpt(0.5)
variables = [0.3, 2.5]
variables_gd = [variables]
for i in range(20):
    variables = optimizer.step(getCost, variables)
    variables_gd.append(variables)
    print('Cost - step {:5d}: {: .7f} | Variable values: [{: .5f},{: .5f}]'
        .format(i+1, getCost(variables), variables[0], variables[1]) )
```

Command

```
pip3 install nump

python3 quantum_eigen_solver.py
```

Output

```
Cost - step     1:  0.3269168 | Variable values: [ 0.95937, 1.49547]
Cost - step     2:  0.0461959 | Variable values: [ 1.20723, 1.11786]
Cost - step     3:  0.0065278 | Variable values: [ 1.30040, 0.97591]
Cost - step     4:  0.0009224 | Variable values: [ 1.33543, 0.92255]
Cost - step     5:  0.0001303 | Variable values: [ 1.34859, 0.90250]
Cost - step     6:  0.0000184 | Variable values: [ 1.35354, 0.89496]
Cost - step     7:  0.0000026 | Variable values: [ 1.35540, 0.89212]
Cost - step     8:  0.0000004 | Variable values: [ 1.35610, 0.89106]
Cost - step     9:  0.0000001 | Variable values: [ 1.35636, 0.89066]
Cost - step    10:  0.0000000 | Variable values: [ 1.35646, 0.89051]
Cost - step    11:  0.0000000 | Variable values: [ 1.35650, 0.89045]
Cost - step    12:  0.0000000 | Variable values: [ 1.35651, 0.89043]
Cost - step    13:  0.0000000 | Variable values: [ 1.35652, 0.89042]
Cost - step    14:  0.0000000 | Variable values: [ 1.35652, 0.89042]
Cost - step    15:  0.0000000 | Variable values: [ 1.35652, 0.89042]
Cost - step    16:  0.0000000 | Variable values: [ 1.35652, 0.89042]
Cost - step    17:  0.0000000 | Variable values: [ 1.35652, 0.89041]
Cost - step    18:  0.0000000 | Variable values: [ 1.35652, 0.89041]
Cost - step    19:  0.0000000 | Variable values: [ 1.35652, 0.89041]
Cost - step    20:  0.0000000 | Variable values: [ 1.35652, 0.89041]
```

Quantum Parallelism

Quantum parallelism is based on the quantum memory register concept. The quantum memory register can exist in a superposition state, which helps in parallelization. If m is the number of quantum bits in the register, the possible states is 2m. A single quantum operation is equivalent to the exponential number of classical operations. Quantum parallel components are modeled as arguments to a function executed on the register.

Tip   Quantum parallelism is based on the superposition of the quantum states in the memory register.

Summary

In this chapter, we looked at techniques related to quantum AI algorithms such as quantum probability, quantum walks, quantum search, quantum deep learning, and quantum parallelism.

CHAPTER 12

Quantum Solutions

Introduction

"To me quantum computation is a new and deeper and better way to understand the laws of physics, and hence understanding physical reality as a whole."

—David Deutsch

This chapter covers the topic of quantum solutions. You will see how quantum AI algorithms are implemented in different real-life solutions. The solutions covered in this chapter are the traveling salesman problem, scheduling, fraud detection, and distribution solution.

Initial Setup

You need to set up Python 3.5 to run the code samples in this chapter. You can download it from `https://www.python.org/downloads/release/python-350/`.

Traveling Salesman Problem

The traveling salesman problem (Figure 12-1) is an optimization problem in the area of computer science. This problem is part of a set of NP-hard problems. This problem is related to a sales executive's itinerary that consists of N cities. The sales executive can visit a city only once and only go back to the starting city after visiting the other cities. This context of visiting only one city is referred to as the Hamiltonian cycle. The travel between the cities has a cost associated. The sales executive tries to minimize the cost of the travel for the complete itinerary.

© Bhagvan Kommadi 2020
B. Kommadi, *Quantum Computing Solutions*, https://doi.org/10.1007/978-1-4842-6516-1_12

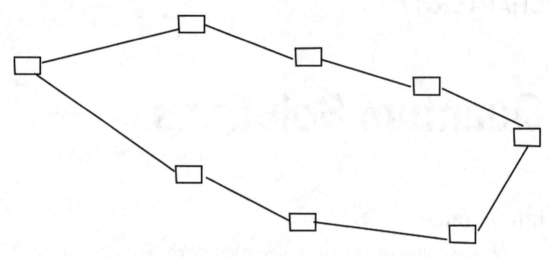

Figure 12-1. *Traveling salesman problem*

Jeongho Bang proposed a quantum technique for traveling salesman problems using the Grover search algorithm. The traveling salesman travels between multiple points, traversing multiple cities along the way. The solution consists of evaluating all the possible route combinations with the associated costs. The optimal route is based on cost, distance, and time parameters. The other parameters are related to the transport type.

This problem is exponential. In exponential problems, a single-point addition increases the complexity. A classical computation framework cannot solve the problem as the number of cities keeps increasing. A quantum computation model helps by using quantum superposition to solve the traveling salesman problem. Quantum computers use the quantum annealing technique.

Tip   The traveling salesman problem is an NP-hard problem because it is complex to find the lowest cost path that connects all the points on the graph. See Figure 12-2.

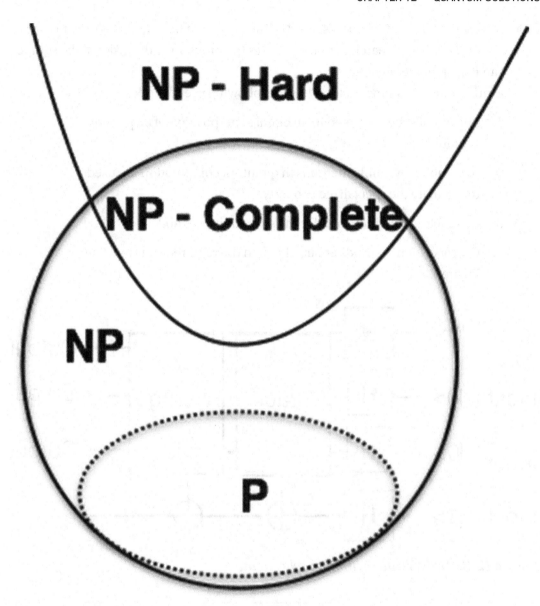

Figure 12-2. *Nondeterministic: polynomial time (NP-hard)*

Scheduling

Scheduling jobs or tasks on a group of resources is a real-life computation challenge. As the number of constraints and resources increases, the problem becomes NP-hard. Linear programming methods are used to solve scheduling problems using classical computation frameworks. Classical computational methods solve the problem in the order of time complexity O(P^Q), where P is the number of tasks and Q is the number of resources. The Grover search algorithm solves the problem in the order of the square root of an R-searchable set of items.

The quantum scheduling method has the following pseudocode:

1. The initial state is created in an equal superposition of all possible solutions.

2. Arithmetic reversible methods in quantum computation are used to calculate the specialized property.

3. An oracle circuit is created for the scheduling method.

4. Grover search methods are used to find the required solution. See Figure 12-3.

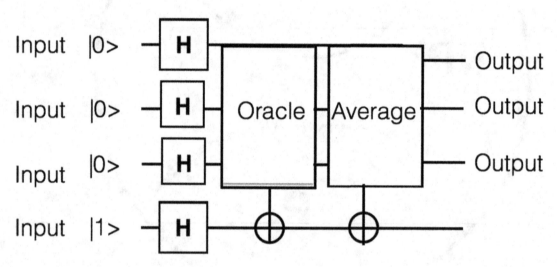

Figure 12-3. *Grover's algorithm: oracle circuit*

Tip   The NP-hard problem is a nondeterministic polynomial-time hardness problem.

Now let's look at the fraud detection solution.

Fraud Detection

When an intruder is making a fraudulent transaction in a system, this typically creates an outlier in the data analysis. This is observed in big data analysis where outliers crop up unexpectedly. The detection of abnormality happens when the data model is trained using machine learning methods. This is observed in multiple areas such as medical insights, data analysis, cleaning, and monitoring. See Figure 12-4.

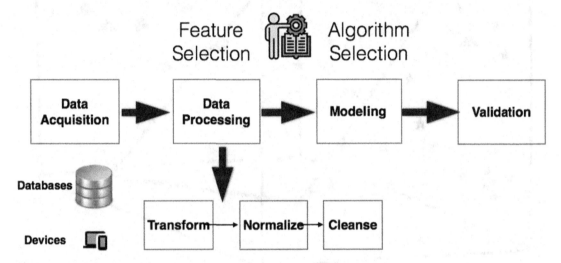

Figure 12-4. *Data processing*

To detect the abnormality, we can analyze the data and find the variation from normal data patterns. Quantum algorithms are used in data analysis to detect abnormalities. Quantum support vector machines and kernel principal component analysis are the two quantum techniques used for fraud detection. These techniques help in finding new quantum observations in the area of quantum security, quantum data transfer, and quantum verification. See Figure 12-5.

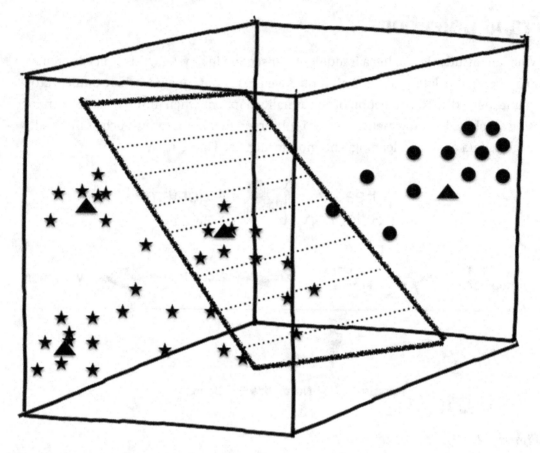

Figure 12-5. *Quantum machine learning*

Tip   Fraud is "the crime of getting money by deceiving people," according to the Cambridge Dictionary.

Distribution Solutions

Distribution planning and optimization problems consist of multi-depot logistics, delivery planning, vehicle routing, logistics integration, consolidation of transport vehicles, single-point and multipoint delivery drop-offs, and warehouse planning and returns. Classical techniques are used for distribution planning and vehicle routing. See Figure 12-6.

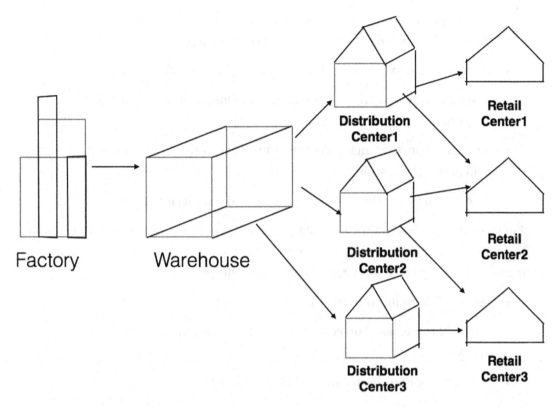

Figure 12-6. *Distribution planning*

Distribution planning covers the schedule route process, planning of resources, vehicle routing and allocation of resources, reporting, what-if analysis, and reoptimization. The quantum algorithms help create transport schedules. These algorithms help in improving the efficiency and moving the enterprises away from silo planning methods.

The following are the constraints in the distribution planning:

- Variables are defined for tracking the cost of every schedule (direct, opportunity, total, reverse opportunity).

- Cost-tracking variables are calculated for every queue (truck schedule).

- The speed of the truck is assumed to calculate the time required to travel from the source to the destination.

- The order delivery time is an important factor for choosing the schedule.

- The opportunity cost is defined as the fraction of the direct cost. This is related to the underutilized capacity of the truck.

- The reverse opportunity cost is a penalty when a truck is oversized.

- For every schedule, the longest time taken for a run is the most important factor for optimization.

- For every run, the truck picks the orders based on the capacity and will return to the source.

- For optimization, the number of trucks is an important factor.

For a given day, based on the booking information between multiple source points and multiple destinations, identify the optimal route for the delivery of goods. The solution is expected to define the following items as output:

- All possible route combinations with associated cost

- Optimal route based on cost (distance, time, minimum number of trucks)

- Alternate options by accumulating delivery date

The problem is purely exponential, meaning for every one-point increase of the source/destination, the combinations to be evaluated increase exponentially. For instance, with three source points, there exists eight distinct combinations to evaluate, and an increase in the source point by 1 doubles the distinct combinations to 16.

While classical computers can solve the early stages, with increased source/destination points, it will gradually degrade in performance and end up breaking. Quantum computers, on the other hand, can leverage the property of quantum superposition to give a constant and better performance.

In the vehicle routing optimization, the goal is to find the optimal routes for multiple vehicles visiting a set of locations. When there's only one vehicle, it reduces to the traveling salesman problem. Optimal routes are the routes with the least total distance. However, if there are no other constraints, the optimal solution is to assign just one vehicle to visit all locations and find the shortest route for that vehicle.

A better way to define optimal routes is to minimize the length of the longest single route among all vehicles. This is the right definition if the goal is to complete all deliveries as soon as possible. We can generalize the vehicle routing problem by adding constraints on the vehicles, including the following:

- *Capacity constraints*: The vehicles need to pick up items at each location they visit but have a maximum carrying capacity.

The goal of the distribution planning solution is to minimize the longest single route for distribution vehicles. Imagine a company that needs to visit its customers in a city made up of identical rectangular blocks.

To solve this problem, you need to create a distance dimension, which computes the cumulative distance traveled by each vehicle along its route. You can then set a cost proportional to the maximum total distance along each route. Routing programs use dimensions to keep track of quantities that accumulate over a vehicle's route. The distance matrix is calculated with source and destination pairs. Pickup point and delivery point input data is provided for the solution to work.

The quantum annealing technique (see Figure 12-7) is used for distribution planning problems. Data is passed to and from the program using JSON files. The quantum annealing process currently supports the optimal vehicle routing problems without time windows. Delivery locations are currently accepted as miles from the depot, which is assumed to be at the coordinates (0,0).

The quantum annealing process is similar to simulated annealing. The thermal activation in the simulated annealing is replaced by the quantum tunneling. This technique is used for computing the minima and maxima for real-life problems such as distribution planning.

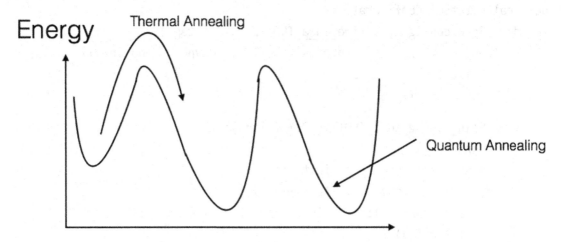

Figure 12-7. *Quantum annealing*

The coordinates are specified in the data.json file. The shortest route distance is the output. The truck's information is in trucks.json. The pickup and delivery information is in pickup_deliveries.json.

Tip   The distribution solution is related to the delivery of goods from the manufacturer to the warehouse and from the warehouse to the store.

Let's look at the quantum annealing technique applied to distribution planning.

Code Sample

```python
import  time
import  math
import  numpy  as  np
import  os
import  random

import json
import argparse

def  distance ( point1 ,  point2 ):
    return  math.sqrt (( point1 [ 1 ] - point2 [ 1 ]) ** 2 +
    ( point1 [ 0 ] - point2 [ 0 ]) ** 2 )

def  calcuatetSpinConfiguration():
    def  Spin_config_at_a_time_in_a_TROTTER_DIM ( tag ):
        config  =  list ( - np.ones ( NCITY ,  dtype  =  np.int ))
        config [ tag ]  =  1
        return  config

    def  Spin_config_in_a_TROTTER_DIM ( Tag ):
        spin  =  []
        spin.append( Config_at_init_time )
        for  i  in  range ( TOTAL_TIME - 1 ):
            spin.append(list( Spin_config_at_a_time_in_a_TROTTER_DIM
            ( Tag[i])) )
        return  spin
```

```
    Spin  =  []
    for  i  in  range ( TROTTER_DIM ):
        Tag  =  np.arange(1,NCITY )
        np.random.shuffle( Tag )
        Spin.append( Spin_config_in_a_TROTTER_DIM ( Tag ))
    return  Spin

def  calculateShortestRoute ( Config,max_distance ):
    Length  =  []
    for  i  in  range ( TROTTER_DIM ):
        Route  =  []
        for  j  in  range ( TOTAL_TIME ):
            Route.append ( Config [ i ] [ j ] . index ( 1 ))
        Length.append ( calculateTotaldistance ( Route,max_distance ))

    min_Tro_dim  =  np . argmin ( Length )
    Best_Route  =  []
    for  i  in  range ( TOTAL_TIME ):
        Best_Route.append ( Config [ min_Tro_dim ] [ i ] . index ( 1 ))
    return  Best_Route

def  calculateTotaldistance ( route, max_distance):
    Total_distance  =  0
    for  i  in  range(TOTAL_TIME):
        Total_distance  +=  distance(POINT[route[i]], POINT[route[( i + 1)
        % TOTAL_TIME]])/max_distance
    return  Total_distance

def  calculateRealTotaldistance ( Route ):
    Total_distance  =  0
    for  i  in  range(TOTAL_TIME):
        Total_distance  +=  distance(POINT [ Route [ i ]], POINT[Route[( i
        + 1 ) % TOTAL_TIME]] )
    return  Total_distance

def  calculateRealdistance ( Route ):
    Total_distance  =  0
    for  i  in  range(len(Route)):
```

```
        if i < len(Route)-1:
            Total_distance  +=  distance(POINT [ Route [ i ]],
            POINT[Route[( i + 1 )]] )
    return  Total_distance

def   moveQuantumMonteCarlo( config ,  Ann_para ):
    c  =  np.random.randint ( 0 , TROTTER_DIM )
    a_  =  list(range(1,TOTAL_TIME ))
    a  =  np.random.choice ( a_ )
    a_.remove( a )
    b  =  np.random.choice (a_ )

    p  =  config [c][a].index(1)
    q  =  config [c][b].index(1)

    delta_cost  =  delta_costc  =  delta_costq_1  =
    delta_costq_2  =  delta_costq_3  =  delta_costq_4  =  0

    for  j  in  range ( NCITY ):
        l_p_j = distance (POINT [p],  POINT [j]) / max_distance
        l_q_j = distance (POINT [q],  POINT [j]) / max_distance
        delta_costc  +=  2*(- l_p_j * config[c][a][p] - l_q_j * config[c]
        [a][q]) * (config[c][a-1][j] + config[c][(a+1)%TOTAL_TIME][j]) +
        2*(-l_p_j * config[c][b][p] - l_q_j*config[c][b][q])*(config[c]
        [b-1][j]+config[c][(b+1)%TOTAL_TIME][j])

    para  =  ( 1 / BETA ) * math.log( math.cosh ( BETA * Ann_para /
    TROTTER_DIM ) / math.sinh ( BETA * Ann_para / TROTTER_DIM ))
    delta_costq_1  =  config [ c ] [ a ] [ p ] * ( config [( c - 1 )%
    TROTTER_DIM ] [ a ] [ p ] + config [( c + 1 ) % TROTTER_DIM ] [ a ] [ p ])
    delta_costq_2  =  config [ c ] [ a ] [ q ] * ( config [( c - 1 ) %
    TROTTER_DIM] [ a ] [ q ] + config [( c + 1 ) % TROTTER_DIM] [ a ] [ q ])
    delta_costq_3  =  config [ c ] [ b ] [ p ] * ( config [( c - 1 ) %
    TROTTER_DIM ] [ b ] [ p ] + config [( c + 1 ) % TROTTER_DIM ] [ b ] [ p ])
    delta_costq_4  =  config [ c ] [ b ] [ q ] *( config [( c - 1 ) %
    TROTTER_DIM ] [ b ] [ q ] + config [( c + 1 ) % TROTTER_DIM ] [ b ] [ q ])
```

```
    delta_cost  =  delta_costc / TROTTER_DIM + para * ( delta_costq_1 +
    delta_costq_2 + delta_costq_3 + delta_costq_4 )

    if  delta_cost  <=  0 :
        config [ c ] [ a ] [ p ] *= -1
        config [ c ] [ a ] [ q ] *= -1
        config [ c ] [ b ] [ p ] *= -1
        config [ c ] [ b ] [ q ] *= -1
    elif  np . random . random () < np . exp ( - BETA * delta_cost ):
        config [ c ] [ a ] [ p ] *= - 1
        config [ c ] [ a ] [ q ] *= - 1
        config [ c ] [ b ] [ p ] *= - 1
        config [ c ] [ b ] [ q ] *= - 1

    return  config

def  verifyRouteConfiguration(routes,pick_deliveries):
    for i in range(len(routes)):
        route = routes[i]
        for j in range(len(route)):
            route_index = route[j]
            #print(route_index)
            for k in range(len(pick_deliveries)):
                pick_delivery = pick_deliveries[k]
                pick = pick_delivery["pickup"]
                delivery = pick_delivery["delivery"]
                if route_index == delivery and route_index !=0:
                    check = verifyPick(pick,delivery,route)
                    if not check:
                        return False

    return True
def  verifyPick(pick_index,delivery_index, route):
    picked = False
    delivery_passed = False
```

```
    for i in range(len(route)):
        route_i = route[i]
        if pick_index == route_i:
            picked = True
        if delivery_index == route_i and picked:
            return True
    return False

if __name__ == '__main__' :

    parser = argparse.ArgumentParser()
    parser.add_argument("-i", "--inputs", type=argparse.FileType('r'),
                  help="inputs as a json file")
    parser.add_argument("-d", "--distances", type=argparse.FileType('r'),
                  help="distances as a json file")
    parser.add_argument("-t", "--trucks", type=argparse.FileType('r'),
                  help="trucks as a json file")
    parser.add_argument("-p", "--pickups", type=argparse.FileType('r'),
                  help="pickups as a json file")

    args = parser.parse_args()

    REDUC_PARA = 0.99

    inputs = json.load(args.inputs)
    distances = json.load(args.distances)
    trucks = json.load(args.trucks)
    pickup_deliveries = json.load(args.pickups)

    TROTTER_DIM = int ( inputs["trotter_dimension"])
    ANN_PARA = float ( inputs["initial_annealing"])
    ANN_STEP = int ( inputs["annealing_step"])
    MC_STEP = int ( inputs["montecarlo_step"])
    BETA = float ( inputs["inverse_temperature"])

    NCITY = len( distances )
    POINT = [[0]*2]*NCITY
    TOTAL_TIME = NCITY
```

```python
num_trucks = len(trucks)

for  i  in  range ( NCITY ):
    distance_node = distances.pop()

    node = [0]*2
    node[0] = float(distance_node["x"])
    node[1] = float(distance_node["y"])

    POINT [NCITY-1-i]  =  node

max_distance  =  0
for  i  in  range ( NCITY ):
    for  j  in  range ( NCITY ):
        node_distance = distance ( POINT [ i ],  POINT [ j ])
        if  max_distance  <=  distance ( POINT [ i ],  POINT [ j ]):
            max_distance  =  distance ( POINT [ i ],  POINT [ j ])
Config_at_init_time  =  list( -np.ones( NCITY,dtype = np.int ))
Config_at_init_time [ 0 ]  =  1

print("starting the annealing process ...")
t0  =  time.clock ()

np.random.seed(100)
spin  =  calcuatetSpinConfiguration ()
LengthList  =  []
truck_routes = []
check = False
for  t  in  range ( ANN_STEP ):
    for  i  in  range ( MC_STEP ):
        con  =  moveQuantumMonteCarlo( spin ,  ANN_PARA )
        rou  =  calculateShortestRoute( con,max_distance )
        length  =  calculateRealTotaldistance ( rou )
        nprou = np.array(rou)
        truck_routes = np.split(nprou,num_trucks)
        check = verifyRouteConfiguration(truck_routes,pickup_deliveries)
    LengthList .append ( length )
```

```
        print("No: Step: {}, Annealing process Parameter: {}" . format ( t
        + 1 , ANN_PARA ))
        ANN_PARA  *=  REDUC_PARA
    check = verifyRouteConfiguration(truck_routes,pickup_deliveries)
    if check:
        truck = 0;
        for route in truck_routes:
            route_distance = calculateRealdistance(route)
            print("Route for truck",truck," ",route, "distance ",route_
            distance)
            truck = truck +1
    Route  =  calculateShortestRoute( spin,max_distance )
    Total_Length  =  calculateRealdistance ( Route )
    Elapsed_time  =  time.clock () - t0
    print("SHORTEST ROUTE : {}" . format ( Route ) )
    print("SHORTEST DISTANCE: {}" . format ( Total_Length ))
    print("PROCESSING TIME: s {}" . format ( Elapsed_time ))
```

Command

```
pip3 install nump
```

```
python3 quantum_annealer.py -i input.json -d data.json -t trucks.json -p
pickup_deliveries.json
```

Input

The input file has the following properties:

- *trotter_dimension*: Dimension of spaced classical spins

- *initial_annealing*: Initial value of annealing factor

- *annealing_step*: Number of annealing steps

- *montecarlo_step*: Number of steps in a Monte Carlo simulation

- *inverse_temperature*: Initial reverse temperature

The sample input JSON file is shown here:

input.json

```
{

    "trotter_dimension": 11,
    "initial_annealing": 1,
    "annealing_step": 1,
    "montecarlo_step": 12000,
    "inverse_temperature":39
}
```

data.json

The data file has the properties mentioned here:

```
id : order  identifier
quantity : order quantity
x :  x coordinate
y : y coordinate
```

```
[{"id": 0,"x": 15.048516730800019, "y": 20.81882158337833},
 {"id": 1,"x": 21.47333104880339, "y": 27.44976598047904},
 {"id": 2,"x": 8.733804547248988, "y": 5.190413472062651},
 {"id": 3,"x": 32.984718431911894, "y": 19.850028824298047},
 {"id": 4,"x": 20.482480291595763, "y": 1.3540523373289077},
 {"id": 5,"x": 28.030738187870487, "y": 14.783008504999733},
 {"id": 6,"x": 14.594496509157707, "y": 2.2840975315280936},
 {"id": 7,"x": 14.988697645586926, "y": 29.21707544497128},
 {"id": 8,"x": 5.553788771696674, "y": 22.21852944451273},
 {"id": 9,"x": 31.95387042131473, "y": 9.878857190484847},
 {"id": 10,"x": 15.52849215384657, "y": 10.543857553552229},
 {"id": 11,"x": 28.837242557836202, "y": 14.368927860603366},
 {"id": 12,"x": 22.776355118030853, "y": 26.698862282083887},
 {"id": 13,"x": 13.377480255985654, "y": 24.833880312965892},
 {"id": 14,"x": 26.031165662014942, "y": 16.32287381266326},
 {"id": 15,"x": 22.609038273840298, "y": 18.014879226920108},
```

    {"id": 16,"x": 32.28095220079001, "y": 34.223176908480426},
    {"id": 17,"x": 0.4255279487934993, "y": 26.28113259271511},
    {"id": 18,"x": 20.032208383820727, "y": 17.778060168894424},
    {"id": 19,"x": 10.227058813719593, "y": 5.158935686180674}]

pickup_deliveries.json

The pickup_deliveries data file has the following properties:

```
id :    order  identifier
pickup :    pickup point index
delivery :   delivery point index
```

[{"pickup":0, "delivery":13,"quantity": 4},
 {"pickup":13, "delivery":18,"quantity": 5},
 {"pickup":18, "delivery":12,"quantity": 5},
 {"pickup":12, "delivery":1,"quantity": 5},
 {"pickup":16, "delivery":3,"quantity": 5},
 {"pickup":3, "delivery":15,"quantity": 5},
 {"pickup":15,"delivery":14,"quantity": 5},
 {"pickup":14,"delivery":11,"quantity": 5},
 {"pickup":9,"delivery":5,"quantity": 5},
 {"pickup":5,"delivery":4,"quantity": 5},
 {"pickup":4,"delivery":6,"quantity": 5},
 {"pickup":6,"delivery":2,"quantity": 5},
 {"pickup":19,"delivery":10,"quantity": 5},
 {"pickup":10,"delivery":17,"quantity": 5},
 {"pickup":17,"delivery":7,"quantity": 5},
 {"pickup":7,"delivery":8,"quantity": 5}
]

trucks.json

The trucks data file has the following properties:

```
id : truck identifier
capacity: truck capacity
```

[{"id": 0,"capacity": 11185},
 {"id": 1,"capacity": 11195},

```
{"id": 2,"capacity": 11195},
{"id": 3,"capacity": 11195}
]
```

Output

```
t0  =  time.clock ()
No: Step: 1, Annealing process Parameter: 1.0
Route for truck 0    [ 0 13 18 12   1] distance  24.885118421529263
Route for truck 1    [16   3 15 14 11] distance  32.16401175641577
Route for truck 2    [9 5 4 6 2] distance  34.18795492132374
Route for truck 3    [19 10 17   7   8] distance  55.9719642261916
quantum_annealer.py:225: DeprecationWarning: time.clock has been deprecated
in Python 3.3 and will be removed from Python 3.8: use time.perf_counter or
time.process_time instead
  Elapsed_time  =  time.clock () - t0
SHORTEST ROUTE : [0, 13, 18, 12, 1, 16, 3, 15, 14, 11, 9, 5, 4, 6, 2, 19,
10, 17, 7, 8]
SHORTEST DISTANCE: 166.92311095246916
PROCESSING TIME: s 22.243867
```

Summary

In this chapter, we looked at the quantum solutions in which different quantum techniques are applied. The distribution planning solution was presented in detail with code samples.

CHAPTER 13

Evolving Quantum Solutions

Introduction

"Quantum mechanics broke the mold of the previous framework, classical mechanics, by establishing that the predictions of science are necessarily probabilistic."

—Brian Greene

This chapter covers the topic of evolving quantum solutions. You will see how quantum AI algorithms are implemented in different real-life solutions. The solutions covered in this chapter are related to quantum annealing, quantum key distribution, and quantum teleportation.

Initial Setup

You need to set up Python 3.5 to run the code samples in this chapter. You can download it from `https://www.python.org/downloads/release/python-350/`.

Quantum Annealing

The quantum annealing process is related to computing real-life problems with low-energy solutions. Quantum bits represent the energy state of the superconducting loops that have low value. The quantum processing unit consists of the superconducting loops.

© Bhagvan Kommadi 2020
B. Kommadi, *Quantum Computing Solutions*, https://doi.org/10.1007/978-1-4842-6516-1_13

Every quantum bit reaches the superposition state after the quantum annealing process is over. The superposition state value can be 0 or 1. During the quantum annealing process, energy moves from the minimum value to a higher value, creating a valley, and the energy diagram looks like a double well with two valleys (see Figure 13-1).

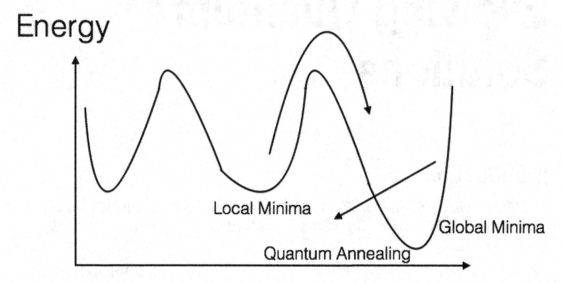

Figure 13-1. *Quantum annealing*

The probability of the quantum bit having a zero or one state is the same. This can be modified by creating the external magnetic field to the quantum bit. The double well with two valleys is created by the field in inverse form. The bias is related to the quantitative value of the external magnetic field. A coupler is used to connect the quantum bit to the bias. (See Figure 13-2.)

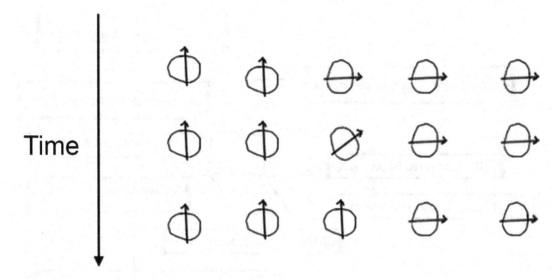

Figure 13-2. *Qubit wise controlled field*

Every quantum bit has a bias. The quantum bits are impacted with couplers. Bias and coupler values can be selected by the user for every problem. The energy landscape consists of the biases and couplings in the quantum annealing process.

Tip   Quantum annealing was created by Tadashi Kadowaki and Hidetoshi Nishimori in 1998.

Quantum Key Distribution

Quantum key distribution (see Figure 13-3) is related to long-term security in an enterprise. It is based on quantum cryptography, which is based on quantum physics. It helps in the secure distribution of encrypted symmetric keys. Quantum particles are sent across the optical network. These particles are photons.

Quantum key distribution is the distribution and sharing of secret keys that are based on cryptographic techniques and protocols. The secret keys are private, and the keys are exchanged between the two parties. The quantum characteristics are not valid if the keys have been tampered with. The enterprise security solutions are based on open quantum key distribution software. These security techniques are in the research stage at this time. Security attacks on a system that use quantum key distribution software are a special research area.

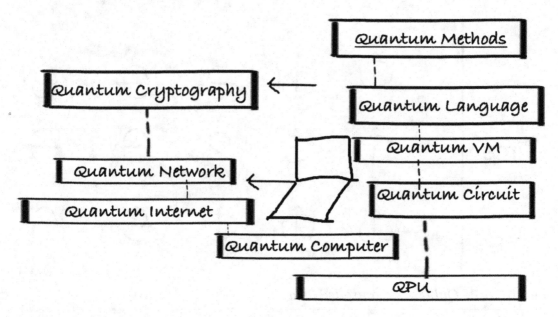

Figure 13-3. *Quantum cryptography*

The quantum cryptographic techniques used in research are based on complete hybrid and high-speed resistant methods. The following methods such as McEliece, NTRU, and lattice-based encryption are still in the research stage for security attacks using quantum methods. Quantum methods are based on the hash, code, lattice, and multivariate quadratic algorithms. See Figure 13-4.

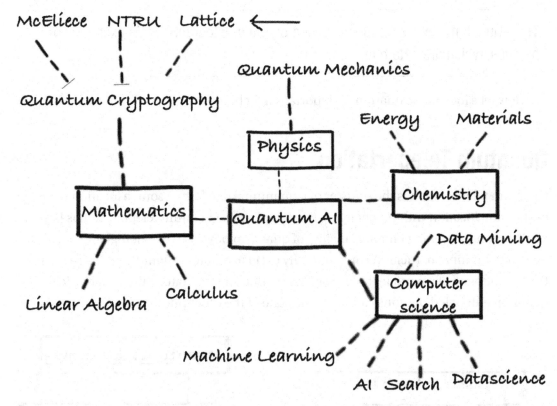

Figure 13-4. *Quantum cryptographic techniques*

Quantum key distribution is based on a fiber or free-space quantum channel. The quantum channel is used to send the quantum state of light from the source transmitter to the destination receiver. In this case, the quantum channel is not secured. The quantum communication link between two parties is secured by authentication. The quantum key exchange protocol is based on quantum properties.

The emerging quantum Internet is based on quantum cryptographic applications, quantum key distribution, and secure identification. The quantum key distribution uses the power of quantum mechanics. Interference detected as quantum key distribution–based communication is tough to change. This helps to secure the message.

The quantum key distribution methods are based on discrete and continuous variables. For example, Silberhorn and Grangier are quantum key distribution methods. Frodo and Sike are based on quantum key exchange. Signature methods are used by Picnic and Tesla. The quantum cryptographic systems are based on multivariate, elliptic, lattice, isogenies, hash, and hybrid-based signatures.

> **Tip** Quantum key distribution is based on the cryptographic data exchange of photons, which are data bits.

Now let's look at the quantum teleportation technique.

Quantum Teleportation

Quantum teleportation is the transfer of a quantum state from a source location to a destination location without changing the actual state. Quantum teleportation is based on the quantum entanglement principle of quantum physics. The entanglement is related to multiple particles having mutually exclusive states. *Mutually exclusive* means finding the state of a particle that is used for computing the state of the next particle. Two quantum entangled photons can know the state of the other photon. See Figure 13-5.

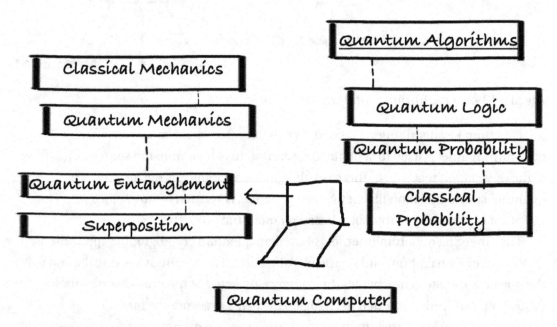

Figure 13-5. *Quantum entanglement*

Each photon can have the four different Bell states mentioned here:

- 00 + 11
- 00 − 11

- 01 + 10

- 01 – 10

The quantum teleportation process starts with the entanglement of a pair of photons in a quantum channel. Photon X is sent by Alice, and photon Y is sent to the receiver Bob. When Alice executes an operation on X, Bob can see the change in the state of Y. Let's say Alice has photon Z whose quantum state is not known. Alice is interested in teleporting into photon Y. Bell measurement is made on photons X and Z. Because X and Y are entangled, Y's quantum state can be turned into the quantum state of Z. Bob can finish teleportation using Alice's Bell measurement. Alice sends the Bell measurement via the classical communication channel. Bob knows how to change Y into the state of Z. See Figure 13-6.

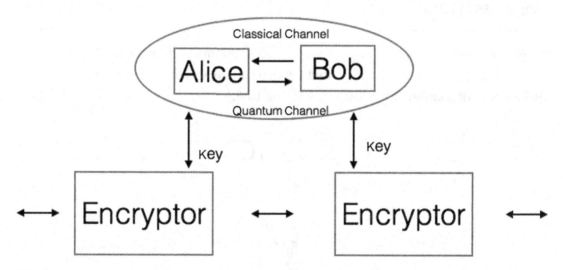

Figure 13-6. *Quantum entanglement*

Let's look at another form of quantum entanglement. The Einstein–Podolsky–Rosen (EPR) pair consists of photons that are in a maximal entangled state. Alice has one of the EPR photons, and Bob has the other photon. In quantum teleportation and quantum entanglement, the particles can be photons, atoms, and leptons of any kind. The particle can have two degrees of freedom. The quantum state of the particle can be encoded into the quantum bit.

The quantum teleportation process is based on the no-cloning theorem. The no-cloning theorem helps to ensure that there is no method to cope for unknown quantum states from one system to another. See Figure 13-7.

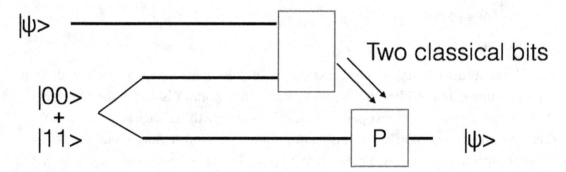

Figure 13-7. *Quantum entanglement*

Tip   Quantum teleportation was discovered by Chaoyang Lu, Jian-Wei Pan, and colleagues in 2015.

Now let's start looking at the implementation of quantum entanglement and quantum teleportation. Let's start with the Bloch sphere, which is used for the visualization of quantum bit vectors. See Figure 13-8.

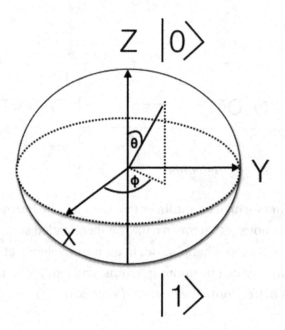

Figure 13-8. *Bloch sphere*

Let's look at the Bloch sphere in code.

Code Sample

```
from qiskit import Aer as qisAer
from qiskit import QuantumCircuit as QC
from qiskit import execute
from matplotlib import pyplot as plot
from qiskit.tools.visualization import plot_bloch_multivector as bloch_
state_vector

quantum_circuit = QC(1, 1)
quantum_circuit.x(0)

simulator = qisAer.get_backend('statevector_simulator')
result = execute(quantum_circuit, backend=simulator).result()
statevector = result.get_statevector()
print('\n statevector is ', statevector)

bloch_state_vector(statevector)
plot.savefig("images/statevector.png")
plot.show()
simulator = qisAer.get_backend('unitary_simulator')
result = execute(quantum_circuit, backend=simulator).result()
unitary_matrix = result.get_unitary()
print('\nunitary matrix is \n', unitary_matrix)
```

Command

```
pip3 install qiskit

python3 BlochSphere.py
```

Output

```
statevector is  [0.+0.j 1.+0.j]

unitary matrix is
 [[0.+0.j 1.+0.j]
 [1.+0.j 0.+0.j]]
```

Figure 13-9 shows the Bloch sphere plot.

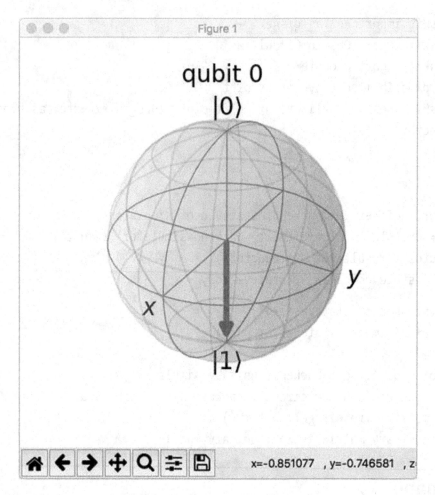

Figure 13-9. *Bloch sphere plot*

Now let's look at quantum entanglement.

Code Sample

```
from qiskit import IBMQ as Quantum_Simulator
from qiskit import QuantumRegister as QReg
from qiskit import ClassicalRegister as CReg
from qiskit import QuantumCircuit
from qiskit import Aer
```

```python
from qiskit import *
from qiskit.tools.monitor import job_monitor as tools_monitor
from qiskit.tools.visualization import plot_histogram
from matplotlib import pyplot as plot

def create_circuit():
    qr = QReg(4)
    cr = CReg(4)

    circuit = QuantumCircuit(qr, cr)

    circuit.h(qr[0])
    circuit.cx(qr[0], qr[1])
    circuit.cx(qr[1], qr[2])

    circuit.measure(qr, cr)

    return circuit

def execute_simulator(circuit, shots=1024):
    print('\n executing the qasm simulator...\n')

    simulator = Aer.get_backend('qasm_simulator')
    result = execute(circuit, backend=simulator, shots=shots).result()

    return result

def execute_machine(circuit, device_name='ibmq_essex', shots=1024):
    print('\nexecuting on the {}...\n'.format(device_name))

    Quantum_Simulator.load_account()

    provider = Quantum_Simulator.get_provider('ibm-q')
    device = provider.get_backend(device_name)

    job = execute(circuit, backend=device, shots=shots)
    tools_monitor(job)

    return job.result()
```

```python
def create_circuit_diagram(circuit, title=''):
    circuit.draw(output='mpl')
    plot.title(title)
    plot.savefig("images/qe_circ.png")
    plot.show()

def create_results(circuit, result, title=''):
    counts = result.get_counts(circuit)
    print(counts)
    plot_histogram(counts)
    plot.title(title)
    plot.savefig("images/qe_res.png")
    plot.show()

if __name__ == '__main__':

    circ = create_circuit()

    res = execute_simulator(circ)

    create_circuit_diagram(circ, '4 quantum bits based qauntum entaglement
    circuit')
    create_results(circ, res, '4 quantum bits based qauntum entaglement')
```

Command

```
pip3 install qiskit

python3 Quantum_Entanglement.py
```

Output

```
executing the qasm simulator...
{'0000': 526, '0111': 498}
```

Figure 13-10 shows quantum entanglement, and Figure 13-11 shows quantum probabilities.

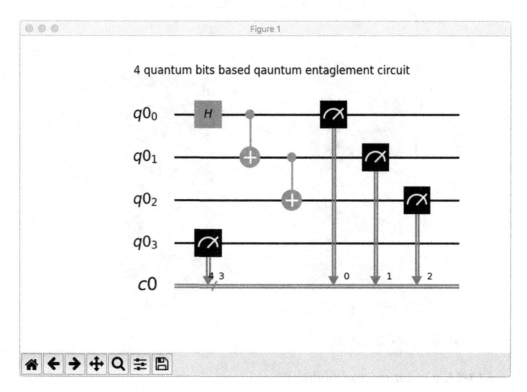

Figure 13-10. *Quantum entanglement*

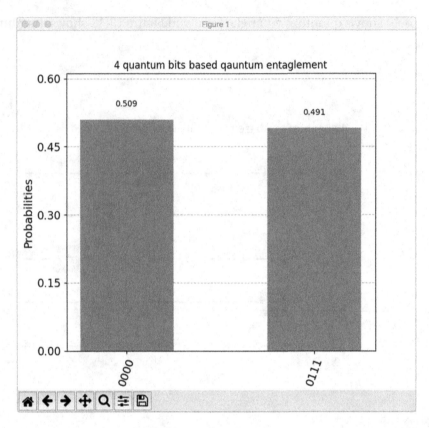

Figure 13-11. *Quantum probabilities*

Now let's look at quantum teleportation.

Code Sample

```
from qiskit import QuantumCircuit as QC
from qiskit import IBMQ as Quantum_Simulator
from qiskit import Aer as Quantum_AER
from qiskit import execute
import matplotlib.pyplot as plot
from qiskit.tools.monitor import job_monitor as tools_monitor
from qiskit.tools.visualization import plot_histogram as histogram

def create_circuit():
    circ = QC(4, 4)
```

```python
    circ.x(0)
    circ.barrier()

    circ.h(1)
    circ.cx(1, 2)
    circ.cx(0, 1)
    circ.h(0)
    circ.barrier()

    circ.measure(0, 0)
    circ.measure(1, 1)
    circ.barrier()

    circ.cx(1, 3)
    circ.cz(0, 3)
    circ.barrier()

    circ.measure(2, 2)

    return circ

def execute_simulator(circ, shots=1024):
    print('\nexecute the qasm simulator...\n')

    sim = Quantum_AER.get_backend('qasm_simulator')
    result = execute(circ, backend=sim, shots=shots).result()

    return result

def execute_machine(circ, device_name='ibmq_essex', shots=1024):
    print('\nexecuting on {}...\n'.format(device_name))

    Quantum_Simulator.load_account()

    provider = Quantum_Simulator.get_provider('ibm-q')
    device = provider.get_backend(device_name)

    job = execute(circ, backend=device, shots=shots)
    tools_monitor(job)

    return job.result()
```

```python
def create_circuit_diagram(circ, title=''):
    circ.draw(output='mpl')
    plot.title(title)
    plot.savefig("images/qt_circ.png")
    plot.show()

def create_results_plot(result, title=''):
    counts = result.get_counts()
    print('\ncounts are : {}\n'.format(counts))
    histogram(counts)
    plot.title(title)
    plot.savefig("images/qt_res.png")
    plot.show()

if __name__ == '__main__':
    circ = create_circuit()

    create_circuit_diagram(circ, ' quantum teleportation circuit')

    result = execute_simulator(circ)

    create_results_plot(result, ' quantum teleportation output')
```

Command

```
pip3 install qiskit

python3 Quantum_Teleportation.py
```

Output

```
execute the qasm simulator...

counts are : {'0100': 230, '0010': 229, '0101': 296, '0011': 269}
```

Figure 13-12 shows quantum teleportation, and Figure 13-13 shows the quantum teleportation results.

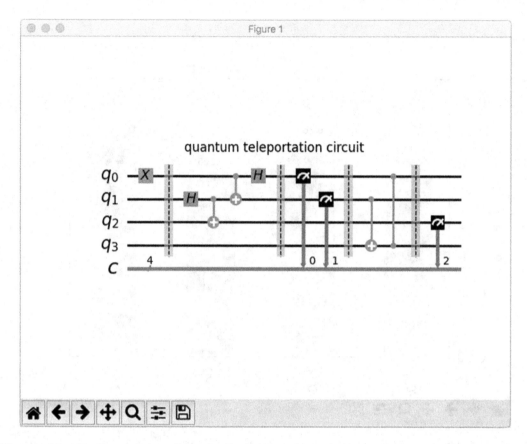

Figure 13-12. *Quantum teleportation*

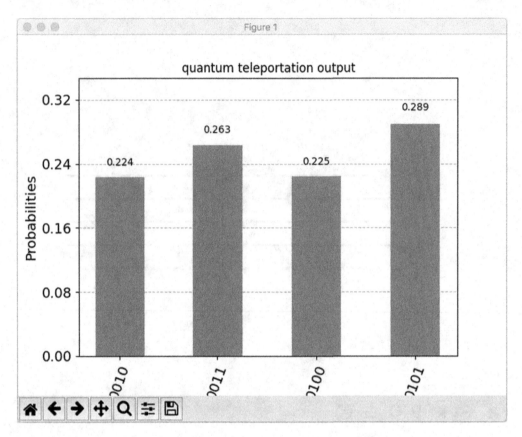

Figure 13-13. *Quantum teleportation results*

Summary

In this chapter, we looked at the quantum techniques used to create quantum solutions in real life. Quantum teleportation was presented in detail with examples.

CHAPTER 14

Next Steps

This chapter covers the next steps. We discuss best practices, case studies, news sources, and the future of quantumware.

Best Practices

The best practices presented in this section are specific to quantum computing frameworks. To start with, a quantum circuit is built by designing it for the specific issue or pain point at hand. The trials are conducted by executing them on different simulators on the server. The results are collated, and stats are gathered at the summary level. The trial results are visualized using software tools.

A software framework related to quantum computing will have package imports, a variables initialization API, a gates modification SDK, a circuit visualizer, a trial simulator, and the results visualizer. As you know, quantum computation is based on quantum bits. Classical computation operates on bits. Quantum bits represent the quantum particle state values. The superimposition state is related to quantum bits having multiple values at the same time. Entanglement is the relation of two quantum bits and their values modification.

Let's look at the quantum computational frameworks.

First, the Strawberry Fields framework is used to create programs related to graphs and networks. The framework has support for machine learning features and is used for real-life chemistry problems. Second, Cirq, which was developed by Google, is used to create quantum circuits and simulate real-life problems on those circuits. Cirq

© Bhagvan Kommadi 2020
B. Kommadi, *Quantum Computing Solutions*, https://doi.org/10.1007/978-1-4842-6516-1_14

has features related to interoperable functions. Another framework, the TensorFlow quantum framework, has features for machine learning. You can easily enhance current TensorFlow apps for quantum-based solutions. The TensorFlow quantum framework uses hybrid quantum classical methods and techniques such as quantum convolutional neural networks. The Microsoft Q# language helps to create applications using the F# and C# libraries. Finally, Qiskit, which is an IBM-based framework, helps to create pipelines and quantum circuits for simulating real-life apps.

Let's look at best practices for creating quantum computing libraries.

- The libraries need to expose APIs that focus on real-life apps using quantum computing models. The functions and methods are created for the algorithms and methods implemented from quantum computing techniques.

- The libraries are designed based on important use cases first to ensure that the SDKs are simple and easy to use. Different components in the SDKs handle different use cases. Examples are created for the usage of each component.

- Libraries need to be created for forward compatibility. The old SDKs are deprecated to avoid issues of compatibility. Shim operations and methods are given for deprecated methods.

- Libraries need to have a goal and purpose for every method and operation.

- Libraries can have classical deterministic methods as methods.

- Libraries need to have methods and operations that are reusable.

- The parameters of the methods and functions need to be selected based on the input and output parameters for an algorithm or a method.

- Libraries need to have namespace names that help the developer to understand the usage of methods, operations, and types.

Case Studies

Intel, IBM, Microsoft, and Google have started research on quantum algorithms. Specifically, IBM, Intel, and Google are focusing on quantum computers. The quantum algorithms research areas include machine learning and artificial intelligence techniques and methods. Voice assistants, digital assistants, and chatbots can use quantum NLP algorithms to improve the efficiency and speed of input processing. Quantum computers can help in pharma drug design, material sciences, molecule modeling, and logistics optimization. The key areas where AI can be blended with quantum machine learning are supervised learning, support vector machines, and neural networks.

Quantum algorithms and quantum computers will be used in complex real-life scenarios. Classical computers will be continued to be used in the areas of process automation, routine task execution, and simple algorithms. Quantum computers will be used in areas such as quantum simulation, optimization, quantum artificial intelligence, and quantum cryptography. They will be used across different verticals such as pharma, cybersecurity, finance, materials science, and telecom.

Now let's consider different applications where quantum computing is being applied.

Applications

These are the areas where apps are being developed:

- Quantum machine learning

- Computational chemistry

- Financial portfolio optimization

- Logistics and scheduling

- Drug design

- Cybersecurity

- Quantum cryptography

- Circuit, software, and system fault simulation

Quantum computers can solve real-life problems in seconds compared to the hours that it would take a classical supercomputer. In the area of financial services, different portfolios can be simulated. In daily life, customers make financial investments, and the investment returns are not predictable. They might need some advice to get better returns. A quantum AI simulator is a working system based on quantum artificial intelligence. Quantum computers can help to simulate customer financial portfolios based on the risk strategy of the customer for predicting returns.

Financial Portfolio Management

Customers have several options when choosing different risk strategies.

- *Protector*: Very low risk

- *Stabilizer*: Low risk

- *Optimizer*: Low to medium risk

- *Creator*: Medium to high risk

- *Maximizer*: High risk

A simulation is based on a mathematical model of a trading situation. The decision variables and the model with assumptions are specified. The initial conditions and alternatives are provided for different scenarios.

Portfolio management is about selecting the best assets to invest in for higher returns based on the risk strategy. It is an NP-hard problem. An NP-hard problem is at least as complex and hard as the hardest nondeterministic polynomial time-based problems. Quantum annealing is the quantum AI technique used to solve this NP-hard problem. Annealing is the equivalent method used to take the glass to the lowest energy state. This is a classical method used in simulated annealing AI techniques. The solution to the hard problem will be at the lowest energy state. As the annealer converges to the lowest energy state, the minima of the problem can be identified.

Quantum annealing is found to solve NP-hard problems, and it has an estimated average of the probabilistic models defined by the Boltzmann distribution. Quantum annealing helps to speed up the solving of NP-hard problems by using quantum phenomenon.

Quantum annealing is about the optimization method to find the minima and maxima of a function. A set of candidate functions is provided to find the optimal minimum and maximum. Quantum annealing is related to the quantum tunneling process. It is a process where particles go through a potential barrier from a high-energy state to a low state. A quantum annealing initial condition is from the superposition state. The system evolves based on the time-dependent Schrodinger equation. This affects the system state amplitude when the time increases. The system's Hamiltonian has resulted from the ground state of the system.

Quantum annealing is used for training Boltzmann machines and quantum deep neural networks. The sampling methods used for finding the probabilistic averages are Markov chain–based Monte Carlo algorithms. The Boltzmann model and equation are the basis of molecular kinetic theory. The most probable velocity, average velocity, and root mean square velocity are calculated by using the Boltzmann distribution.

The trading problem needs to be characterized by defining the goals and business scenarios. The method to solve the problem will be a simulated annealing-based quantum AI algorithm. The deployment will be done in a trading and wealth management firm. Continuous evaluation is done by gathering the solution metrics.

The characteristics of the trading problem are the stock prices, mutual fund prices, time for simulation and prediction, and different alternate paths. The method used in the simulation is quantum annealing. The model will have entities that are stocks and mutual funds. The model variables are the stock and mutual fund prices. The solution will be the strike prices and the assets to be chosen for the investment portfolio. The deployment will be done on the trading and wealth management firm's infrastructure. The environment will include the servers that can handle the required memory, the computing processing power, and the disk space. The system sizing is done to handle the customer portfolios of an order of 10,000 stocks and simulating them for more than 10 years. The continuous evaluation of the solution is done by executing the solution on the servers. The metrics are gathered for the solution including the portfolios selected and the expected returns based on the risk profile.

The quantum AI simulation-based annealer will have a knowledge base, inference engine, data store, and explanation module. The knowledge base will be the historical data of stocks and mutual funds. The inference engine will have rules and the facts for trading and portfolio management. A data store is used for portfolio management. The explanation module helps in measuring the metrics based on the goals, risk strategy, and the initial portfolio selected.

The sampling methods that are Monte Carlo based are used to value stocks and mutual funds in the finance world. The random shifts in the prices of the assets are simulated. The Monte Carlo methods find risk measures in trading and investment strategies. They help in finding path-dependent returns for a risk-based investment strategy. Estimation of the return and risk for a portfolio of stocks, mutual funds, options, and futures is done by using Monte Carlo quantum methods. The volatility helps in identifying the variation and can be computed by the logarithmic returns' standard deviation.

Quantum AI simulation goes through forward direction–based steps in time. The initial asset prices are taken from the user input or from the stock market websites. The quantum AI model consists of the asset price evolution. Stochastic processes or discretization methods are used for solving the problem. The asset prices might have correlations that are not expected due to random data generation. Quasi-random number generators help to replace the complete random number generators. Quasi-random numbers are based on a quasi-random sequence, which is of low discrepancy.

Quantum Monte Carlo methods help in modeling the stock prices using stochastic Brownian motion. This model consists of fractional and large quantities of assets without transaction prices. Short-selling and dividend-less stocks are the assumptions in the model. The value at risk is the goal of the model. It is related to the expected loss of the customer portfolio.

For example, the option price starts from the values between 11.4 and 11.2. The exercise price for the option can be around 15. The maturity of the option is around 1 year. The volatility of the stock price is expected at around 0.2 (20%). The simulations are computed for different volatility values between 0.15 and 0.25. The number of steps within a simulation is around 10,000.

Monte Carlo simulations are based on Markov chains. They are related to quantum walks, and quantum walks are analogous to random walks. Quantum Monte Carlo methods are applied for stock pricing, risk-based investment strategies, return computations, portfolio return calculations, and impact of investment in assets.

The quantum AI simulation helps in forecasting the outcome, which is indeterministic. A random walk with all the states in the data store can be evolved into a trading system with multiple stocks and mutual funds. A probabilistic average from the possible states is the goal of the simulation. The quantum algorithms based on Monte Carlo methods are computed m times to obtain m number of random combinations of the stocks and mutual funds in a customer portfolio.

A quantum AI simulator can execute the simulation for more than 10,000 future scenarios. The scenarios will not have the following key factors:

- Market crashes

- Company-specific events

- Political surprises

- Unexpected interest hikes

The simulation results are not accurate as these key factors might disrupt the stability of the trading markets with a downside of risk and an upside of potential. Quantum AI simulation can handle stocks, mutual funds, and options such as American, European, Asian, LookBack and Bermudan styles. The simulation model will have the following variables:

- Today's stock price
- Exercise price
- Maturity in years
- Risk-free rate
- Volatility
- Number of simulations
- Number of steps

The other parameters that influence the simulation are commodity prices, inflation, and other assets in the trading markets. Mathematically, the simulation will look as follows using the normal distribution:

Stock_Price(t+1) = Stock_Price(t) *Pow(e, (drift+random_value))

drift = average_stock_price - (variance/2)

random_value = (standard_deviation * normal_inverse)

normal_inverse is the inverse of the normal distribution.

The distribution can modeled with Q(x) as the frequency distribution or the density function r(x) (x is continuous).

If there is a real-value function f(X) with the probability distribution for frequency Q(x), the expected value of f(X) is as follows:

E(f(x)) = Σ f(x) Q(x) where Q(x) > 0 and Σ Q(x) = 1

X is discrete.

The f(X) mean value is $(1/n) \Sigma$ f(xi), where i lies between 1 and N.

The investment horizon is typically between 1 and 25 years in the simulation model. The asset's downside and upside price are the inputs. The simulation models the risk-neutral paths of a portfolio with stocks, mutual funds, and options. Bumping is related to changing the sensitivity of the parameters in the model and executing the simulation.

The complicated parts in the quantum AI simulator are knowledge base, inference engine, and explanation modules.

The knowledge base is created by transferring and transforming the domain expertise from different sources to a semantic model. The domain consists of general principles, entities, and events that commonly occur. Entities in the domain might have relationships and underlying structure. The knowledge base captures the tasks that are performed by the operators at different function levels in the domain. The organization model has the function levels. The task model has the operator's tasks in detail. The application model will have information related to the operator's systems. Ontology is a key part of the knowledge base. Static ontology is related to domain entities and their relationships. Dynamic ontology is about the states of the entities and the transformation rules. Epistemic ontology is related to state transformation guidelines. Domain actions, predicates, procedures, object hierarchy, and finite-state tables are part of the model.

A knowledge-based system can be one that predicts the weather, and it has knowledge in meteorology. It can forecast weather with good accuracy. Knowledge engineers who are meteorologists will add their knowledge in the form of facts and rules. Similarly, an Army strategist can come up with military strategies using the knowledge base of the defense. The knowledge base will have a standard set of strategies for attack and defense. See Figure 14-1.

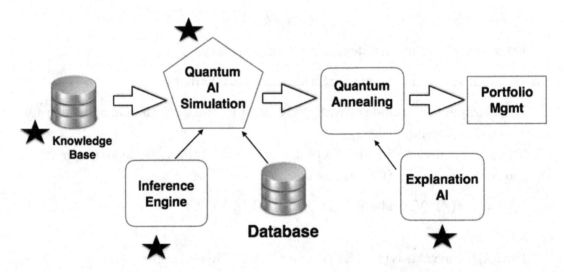

Figure 14-1. *Portfolio management*

The statistical models related to the trading and advisory knowledge base will evolve based on the historical data and the recent trends from the stock market. The market events, objects in the trading domain, observable patterns, trading strategies, and business policies are part of the trading system knowledge base. The knowledge base will have structural, strategic, and support knowledge types. The structural knowledge will have different abstraction levels in the trading domain. Strategic knowledge is related to methods and techniques to solve different problems in the trading world. Support knowledge is for contingencies and the causes of problems in the trading domain.

The inference engine executes the rules in which the knowledge base has to identify the change in the states of the objects and the entity's properties. The engine solves the problems based on the rules in the knowledge base. The inference engine helps in answering the questions that came up during the simulation.

The explanation module has declarative rules for measuring the effectiveness of the solution. The explanation is related to evaluating the system results and identifying the cause by reasoning. The explanation and reasoning are based on different rule types such as causal, trigger, data, hypothesis, and screening rules. The quantum AI simulator identifies the scenarios to be executed based on a set of events. See Figure 14-2.

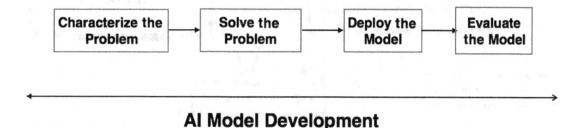

AI Model Development

Figure 14-2. *AI model development*

Moving on from banking, let's look at other verticals such as healthcare. In healthcare, quantum AI can be used to analyze genetic DNA sequences. The size of the genomic sequences is big. Quantum AI can be used for image processing and analysis in bioengineering and telemedicine. Critical diseases such as cancer can be diagnosed by using these techniques. It also helps in drug discovery and research. These techniques provide better forecasting and time-series analysis.

Quantum processing capabilities help in traffic management, sensor data analysis, supply chain planning, procurement decision-making, searching big data, transportation optimization, resource planning, and asset management. Like bits, quantum bits are the computational basis for the quantum computer.

Now let's look at traffic management.

Traffic Management

Traffic modeling in transportation depends on traffic zones, trip tables, highway networks, vehicle capacity, and land-use models. Land-use models consist of trends, land-use and zoning plans, activity areas, and maps. A transportation system has streets, highways, parking, travel times, and speed limits. The transportation network will have traffic volumes and vehicle capacities. Travel patterns identify the mode of use, typical source to destination paths, and the trip characteristics. Wardrop's first principle is based on the network and road user balance. The journey times of all the paths used are equal and less than the unused path by one vehicle. This principle helps in minimizing the average time for a trip. There are other principles and algorithms to balance the road user and network traffic flow. See Figure 14-3.

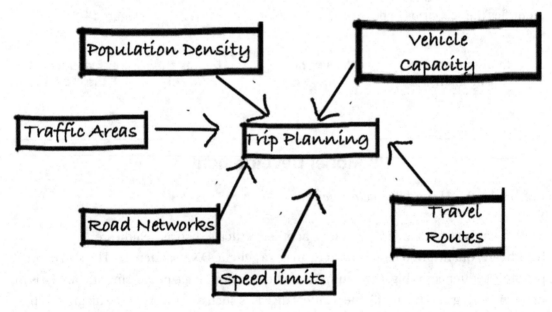

Figure 14-3. *Trip planning*

Trips are generated based on the source, destination, vehicle, and city transportation network. The network has vehicles, cyclists, pedestrians in different time zones, and space all constantly moving. Transportation planning is a continuous activity, and the street network infrastructure keeps getting updated. Traffic modeling is done using techniques such as simulation and machine learning. Traffic control is done using historical data, estimating the number of vehicles during a time period, and managing the flow of the traffic across the network.

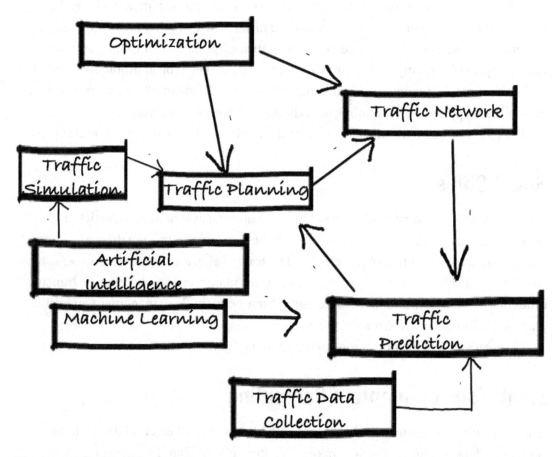

***Figure 14-4.** Traffic planning*

A max flow–min cut approach is used to maximize the flow of the traffic in a transportation network. This approach, based on the maximum flow of traffic from the source to the destination, is the same as the minimum cut to separate the source from the destination. This approach can be applied to a computer network packet flow.

The packet flow through a network can be modeled using the max flow–min cut approach. The network has the information flow through the nodes to reach the destination. The goal is to maximize the number of packets that can flow through a computer network. Similarly, this method can be used to assign students to a dorm based on the condition that every student wants to live in one's own dorm of choice. The goal is to maximize the number of students who get dorms.

In other verticals, quantum algorithms and techniques are used for predictive analytics. In healthcare, the quantum annealing method can help in optimizing the nonadverse effects in protein folding and drug discovery. The other areas are drug design and personalized medicine. In manufacturing, quantum computing techniques can be applied for supply chain optimization problems. The typical problems are related to distribution, production, procurement, and logistics. In the media industry, quantum methods are applied for scheduling advertising and optimizing revenue.

Let's look at the quantum algorithms and techniques implemented in smart cities.

Smart Cities

The problems in smart cities consist of big data handling and require analytics and decisions in real time. A city consists of cars (autonomous and human driven), trucks, autos, buses, and other transport vehicles. The traffic lights are on different streets of the city. The traffic history of each street is available for planning a trip at different transit times on the weekdays and on the weekends. Smart city traffic management and drive planning solutions help citizens plan trips.

Now, let's look at the limitations of quantum computing and algorithms.

Limitations of Quantum Computing

Quantum computing hardware has issues with vibrations and external interactions such as radiation, light, and other factors. Decoherence is related to measuring the state at different points and collapsing the quantum states. The probability of decoherence increases with the number of quantum bits in exponential fashion. Phase errors are induced by using the quantum methods. Classical methods can cause small errors by using error correction methods. Testing is a problem as different copies of the unknown state cannot be created. Superposition is required for quantum computing methods, but a realistic model can be achieved as superposition requires the lowering of the system entropy.

Quantum Computing News Sources

The following are the news sources related to quantum computing:

- *IBM research blog*: `https://www.ibm.com/blogs/research/category/quantcomp/`

- *Microsoft Quantum*: `https://cloudblogs.microsoft.com/quantum/`

- *Waterloo Institute Quantum blog*: `https://uwaterloo.ca/institute-for-quantum-computing/news`

- *ACM SIGARCH*: `https://www.sigarch.org/tag/quantum-computing/`

- *The Qubit Report*: `https://qubitreport.com/`

Now let's now look at quantumware and its future.

The Future of Quantumware

The quantum bits on a quantum processing unit are increasing and have been doubling every year for the past ten years. The goal of different companies such as IBM is to scale the manufacturing process to create quantum processing units of size 10,000. Quantum computers are being used by Google, IBM, Microsoft, and NASA in their research and company IT infrastructure.

According to the CEO of the D-Wave company, Vern Brownell:

> *Google has created what they call the Quantum Artificial Intelligence Lab, where they're exploring using our computer for AI applications or learning applications. And NASA has a whole set of problems that they're investigating, ranging from doing things like looking for exoplanets to [solving] logistic problems and things like that. I'd say within five years, it's going to be a technology that will be very much in use in all sorts of businesses. While business applications within quantum computing are mostly hopeful theories, there's one area where experts agree quantum could be valuable: optimization. Using quantum computing to create a program that "thinks" through how to make business operations faster, smarter and cheaper could revolutionize countless industries.*

Quantum AI is related to learning models inspired by a biological brain and quantum mechanics. These models are based on mathematical operations on quantum bits that are used for solving real-life problems. Distributed systems and parallelized architecture–based systems are used for running these algorithms and models. As a quantum bit can handle multiple states such as 0 and 1, huge problems can be solved in a short time with n qubits.

Heisenberg's uncertainty principle uses information about the particle's momentum and location. Any observation for calculating the location will cause an error in the momentum of the particle. Similarly, any calculation of the momentum causes an error in the location. Quantum AI algorithms are based on the set of the transformation of quantum bits. The computation does not measure the state of the system that causes an error in the information. A single quantum algorithm cannot find all solutions to an NP-complete problem. The structure of the problem in the context is used as a basis for solving the NP-complete problem.

Quantum AI cannot help to change routine office tasks such as reading, writing, and mathematical calculations. A quantum computer will not be useful when reading big data, writing reports, or performing mathematical calculations. The sequential operations in an enterprise will require a classical computer. Complex algorithms will be executed on the quantum computer in an organization going forward.

Classical computers and servers will be used for a standard set of processes that have manual and automated operations. These operations might be related to validation, checking, workflow, communication, notification, data gathering, and data storage. Data streaming and a push model will require standard computers that gather data and stream them to various subscribed parties.

Enterprise Software and Services for Quantum Computing

In enterprise software and services, solutions related to chemical, pharma, and manufacturing are evolving. Drug discovery and the reduction of research and development are the key areas in pharma for applying quantum techniques and methods. Quantum software is being developed to solve real-life optimization problems. The other area of application is autonomous driving. Cybersecurity/cryptography is another area where quantum software and solutions are evolving.

Quantum Machine Learning

Quantum machine learning (QML) is another area where new solutions and products are evolving.

Summary

In this chapter, we looked at the next steps for quantum techniques to be used to create quantum solutions in real life.

Index

A

AI model development, 287
Antiparticles, 3
Approximate algorithms, 125
Approximate combinatorial
     optimization, 125, 126
Approximate optimization
     algorithm, 125, 126
Arithmetic logic unit (ALU), 24

B

Bounded-error quantum |
     polynomial, 137, 138

C

Chain molecule structure, 4, 5
Cirq, 279
Classical clustering techniques, 209, 210
Classical computation, 242, 279
Classical computers, 7, 281, 292
Classical error correction techniques, 76
Classical *vs.* quantum bits, 49, 50
Classical SVM, 185, 186
Classic search algorithm, 235
Classifiers
     code sample, 176, 180–182

command, 177, 183
     output, 177, 183, 184
     quantum, 174, 175
     variational quantum circuits, 177–180
CNOT gate circuit, 71
Coined walk model, 228, 229
Column matrix, 36, 37
Combinatorial optimization,125, 126,
     137, 138
Continuous function, 142
Continuous-time quantum walk, 232–235
Controlled quantum, 71
Controlled-U gate circuit matrix, 72
Control unit (CU), 24
Cross-currency arbitrage, 19
Cryptography
     financial firms, 16
     functioning cryptographic
          systems, 16
     Grover's algorithm, 17
     Heisenberg's uncertainty principle, 17
     models, 16
     quantum computer–based
          algorithm, 17
     quantum key distribution, 16
     Quantum Resistant Ledger, 16
     superposition, 17
Cybersecurity, 19, 20, 281, 292

© Bhagvan Kommadi 2020
B. Kommadi, *Quantum Computing Solutions*, https://doi.org/10.1007/978-1-4842-6516-1

Printed in the United States
By Bookmasters